The German Sturmgeschütze
in World War II
1939-1945

Schiffer Military History
Atglen, PA

Photo Credits

Arndt (1), Bernd (2), BA (42), Carls (2), Doss (18), Eiermann (43), Elser (5), Fleischer (117), Flemming (1), Hoppe (8), Kesczcki (3), MHM (10), Rahne (1), Regeniter (15), Thiede (7), Wetzig (6).

Back cover:

British soldiers examine a shot-up German assault gun, western front, 1944.

Translated from the German by Ed Force

Printed in China.
ISBN: 0-7643-0693-6

This book was originally published under the title, *Die deutschen Sturmgeschütze 1935-1945* by Podzun-Pallas.

We are interested in hearing from authors with book ideas on related topics.

Published by Schiffer Publishing Ltd.
4880 Lower Valley Road
Atglen, PA 19310
Phone: (610) 593-1777
FAX: (610) 593-2002
E-mail: Schifferbk@aol.com
Please visit our web site catalog at www.schifferbooks.com
or write for a free catalog.
This book may be purchased from the publisher.
Please include $3.95 postage.
Try your bookstore first.

Wolfgang Fleischer
with
Richard Eiermann

The German Sturmgeschütze in World War II

1939-1945

A Photo Chronicle

An assault gunner of the "Grossdeutschland" Assault Gun Unit during a pause in combat. To a much greater extent than was the case in the other service arms, these soldiers combined characteristics of the artilleryman, the tank driver and the infantryman—surely a reason for the success of the assault artillery of the German Army in World War II.

CONTENTS

Foreword 7
A new artillery weapon arises 9
Assault artillery—the artillery's armored fist 21
Facing new tasks 29
From assault gun to pursuit tank 73
The cure-all assault gun 97
Fitting into a changed organization 104
The assault gun in battle 107
At the limits of performance 115
The last year of the war 137
Bibliography 160

Contemporary illustration showing the use of an assault gun, 1942.

FOREWORD

The assault gun was a typically German weapon, and not just because of its name. Although already developed in the latter half of the thirties, it ranked among the most successful and characteristic German weapons in World War II. Its imitation by the armies of other countries attested to the correctness of the basic tactical concept, and was at the same time a reference to the success of the assault gun as a weapon. And yet, in the years after World War II, these vehicles disappeared from the scene of the fighting forces, aside from individual developments and a discussion, sometimes carried on vigorously, in the specialist press, about the good and bad points of casemate tanks, for such vehicles were actually what the assault tank amounted to.

Wherein were the reasons for the particular success of the assault gun in the comparatively short period of its existence?

In our search for answers, we should begin by remembering the historical origins of this weapon's development, They go back to the late Middle Ages. At this time, firearms were taking on a growing importance in military activity. Even then, the field commanders were trying to maneuver with the still cumbersome guns, in order to gain advantages on the battlefield and thus gain victories. At first the combination of firepower and mobility succeeded only very imcompletely. As a rule, the guns were too heavy. Lighter guns, such as the leather six-pound cannons used by Colonel Wurmbrandt in Swedish service, were not particularly durable and lacked the necessary firepower. They were utilized, among other times, in 1631 at Leipzig.

As early as 1656, small, mobile guns were an important part of the regiments of the electoral army of Brandenburg. About a hundred years later, the Prussian King Friedrich II, whose infantry had lost much of their original quality in the heavy losses of the three Silesian wars, had his battalions use light three- and six-pound cannons. Since the days of that great Prussian king, the principle has remained: The worse the infantry, the more important the artillery. A greater use of the artillery to support the infantry was opposed for important reasons. Along with the laborious procedure of loading and aiming the front-loading gun in the face of the enemy, there was always the gathering of men, draft horses and materials on the battlefield that prevented their more extensive use. In addition, the insufficient range of 800 to 1360 meters (320-480 m with canister shot) and the lack of armor for protection made them all too sensitive, even against fire from infantry weapons. In principle, these reasons prevailed until the beginning of the 20th century, and were often cited by the critics of a closer link between artillery and infantry.

In World War I (1914-1918), specially designed light guns were used again, with their primary function being the shooting down, with aimed single shots, of machine-gun nests, which had lost none of their oppositional ability despite intensive artillery preparation. This required close cooperation with the infantry. The guns had top be light, which lessened their effect as a weapon, but increased their mobility, which then as before had to be provided by man- or horsepower. Only the steel gun shield offered the crew insufficient protection from the noticeably increased effect of weapons on the battlefield. Thus this weapon too remained a compromise. In the minds of its contemporaries, the concept of assault-assisting firepower was ingrained. There was also a new awareness that, if in future warfare an infantry attack was to be held in check by the much-increased firepower of the defenders' front line, constant support from specialized artillery was needed. From this point there arose the infantry guns that were an integral part of the infantry after 1933 and were used in action by regimental groups.

In the mid-thirties, the decision was made in Germany to build up operative Panzer units. Linked with this was a stronger concentration on the forces and means that were available in limited numbers. The Army's infantry units were threatened with the danger of having to get along in the future without any armored support, which would necessarily have to limit their ability to carry out attacks. In these complex conditions, intensive theoretical discussion, some of it carried on in the specialist press, had to be evaluated. The high point of this disputation was a memorandum from the later Field Marshal Erich von Manstein, who communicated with the Chief of the General Staff and the Commander of the Army in 1935. In it the suggestion was made to utilize the assault support and protection of the infantry, used in World War I, to make use of the progress made in technology in the form of a self-propelled armored gun mount for direct support of the infantry. The development of this weapon, which remained in the province of the artillery, could be completed, despite a series of setbacks, by the time World War II began. Every division was to have a unit of such guns. Their use in the first war years confirmed the correctness of the concept. The need for assault guns grew steadily, though not all needs could be fulfilled.

Experience in action, changing combat conditions, and the further technical development had resulted in a multitude of changes in the organization of structuring and in the basis of combat during the course of the year. The most serious change came about as a result of contact with an overwhelming Russian armored force on the eastern front, the change of the assault gun from an attacking weapon of the infantry to one of the most important and successful of all armored defensive weapons. All efforts to give the assault gun back its original character, such as by arming it with a 10.5 cm assault howitzer, failed in one area—that of antitank use.

In 1943-44 there was no lack of attempts to envision the assault gun as a cure-all to hold the breaking fronts together. The assignment of assault guns to tank and antitank units was one such attempt, surely born of necessity. It gained no success and could not change the fact that the assault gun showed its greatest success as an artillery weapon.

Important encouragement and worthwhile support in the development of this theme were received by the authors from former members of the assault artillery, of whom Dr. Alfred Regeniter, Gerhard Doss and Gerd Albert may be named as representatives. To them, our hearty thanks. Extensive support came from the Military Library of Dresden and the Military Archives in Freiburg. Thanks also to Gerhard Thiede for doing the photographic work, and to Sonja Wetzig, who contributed often and in many ways to the total result.

Freital, autumn 1996
Wolfgang Fleischer
Richard Eiermann

A New Artillery Weapon Arises

A good thing, or even better, a successful thing, always has several fathers. Thus it is no surprise that in the development of the assault gun, several well-known German military men can be named, who, standing in different positions and surely working with differing motives, contributed to that development. For its evaluation it is debatable whether the colonel on the General Staff, Erich von Manstein, should be named as the creator of the new weapon, since he had written, as early as 1935, a memorandum about it, or whether one names another colonel on the General Staff, Walther, who, as the leader of the (8th) department for technical questions, surely had an influence on its development. Without specifically mentioning the name "Sturmgeschütz", the Chief of the General Staff, Generaloberst Ludwig Beck, expressed himself to these problems in a memo of December 30, 1935. Its title was "Considerations on the Heightening of the Attacking Power of the Army". Behind all these documents there was the basic question: How can, in the face of increased weapon effectiveness of defending forces, the infantry's capability to carry out offensive action on the battlefield be maintained? This problem was intensifying because it could already be seen that with the accelerated buildup of tank forces, the infantry divisions of the army had to get along without any armored support. Armored protection of infantry guns, which were being introduced in the 13th companies of the infantry regiments, was also lacking. One of the many answers to the question of how the infantry's penetrating power could be maintained obviously involved the development of special assault guns, which was still based on the idea of horsedrawn infantry assault weapons. Armor protection and engine

The armored self-propelled mount for the 7.5 cm assault gun. The O series used the chassis of the Panzerkampfwagen III (3.7 cm) (Sd.Kfz. 141), Type B, recognizable from the front by the two round servicing ports in the bow of the hull. Five vehicles were built, all of soft iron.

Training of assault artillerymen on O series vehicles by the Artillery Instructional Regiment at the troop training camp in Jüterbog.

power were to help this weapon become an up-to-date technical and tactical development. The higher firepower of infantry advancing with the support of assault guns must, provided they were used properly, necessarily result in a higher penetrating power. The suggestion to develop armored assault guns was basically approved by the Chief of the Army General Staff. In further considerations, Manstein expressed himself in June 1936 on the character and the future use of such assault guns. He wrote: "The assault artillery (whether it takes the form of battle tanks or of armored motorized guns) is rather an auxiliary weapon of the normal infantry division. Its use corresponds to the escort batteries of the last war. . . . " Its main task was the support of the infantry in attack. In defensive action, the assault artillery should act as a part of the division's artillery, for which reason the guns must be suitable for indirect firing at distances up to 7000 meters. This idea was still to be found in the summer of 1940 in the document "The Assault Battery—Possible Uses and Training", but was soon dropped in the course of the war. Noteworthy for this time period was Manstein's assertion: "In the end, it (the assault gun weapon) will be an outstanding means of offensive tank defense, and can perhaps replace the division's antitank unit in this task." This positive concept was to have a considerable effect on the technical and organizational development of the assault gun in the following years.

The very broad spectrum of tasks for assault guns was not essentially limited before the beginning of World War II. Practical experience from intensive troop training was not available. Material prerequisites were also lacking. By September 1939, only five test assault guns on the Panzerkampfwagen III chassis, Type B (Type 2/ZW), were available, their production having

Along with the armored self-propelled mount for the 7.5 cm assault gun, a light armored ammunition transport vehicle (Sd.Kfz. 252), here a pre-series vehicle, recognizable by its road wheels, was utilized. Some of the combat supplies of the 7.5 cm assault gun were carried on this vehicle and its trailer (Sd.Ah. 32).

Loading ammunition for the armored self-propelled mount during training by the Artillery Instructional Regiment.

The gun leader carefully examines the battlefield, so he can direct the armored self-propelled mount for the 7./5 cm assault gun, standing ready in the background, into the most advantageous firing position. The picture was taken during training at the Artillery Instructional Regiment at the troop training camp in Jüterbog.

begun in 1937. They had an armored casemate of cast iron and were thus not to be used on the front lines. The vehicles were used for training purposes at the Assault Artillery School in Jüterbog.

Of the armored self-propelled mount for the 7.5 cm assault gun cannon, Type A, as the laborious designation for the first series-produced Assault Gun III ran, thirty were ordered by the Army Weapons Office (HWA) on the 5/ZW chassis (Panzerkampfwagen III, Type E). The contract could be expanded to forty vehicles, which were delivered by May 1940. The variation of the Panzer III chassis for assault-gun production was naturally accepted with reservations by the Panzer troops. This type was seen, along with Panzerkampfwagen IV, as the main equipment of the Panzer units, a goal whose fulfillment was still far in the future. On May 10, 1940, the day on which the campaign against France began, only 349 Panzer III tanks were available. There were simply no spares. The military requirements for an assault gun, on the other hand, made it necessary to go back to a chassis with the roominess and performance already offered by that of the Panzer III tank. Important details of the requirements were:

1. The chassis had to provide a stable gun platform for a gun whose targets could be aimed at and fired on successfully, preferably directly and within a short time.

2. The effect of the single shot could not be less than that of the infantry guns of World War I. Here the decision was made in favor of a variant of the 7.5 cm KWK 37 L/24, which was already being produced for the Panzerkampfwagen IV and was now mounted on a socket in the hull of the vehicle for use as an assault gun.

3. The mobility of the assault gun should make it possible to follow the infantry, even in the roughest terrain. To do so, sufficient reserves of power were necessary.

In January 1940 a Type A assault gun was delivered. Three more followed in February and another six in March. With these few vehicles, and still under winter conditions, the technical and tactical training of the crews was hastened at the Jüterbog training camp. The first independent assault-gun batteries were established.

Driver training in March 1940. The 300 HP Maybach engine gave the self-propelled mount, which weighed 119.6 tons, a maximum speed of 40 kph. Its off-road capability under unfavorable conditions far exceeded those of the earlier, horsedrawn infantry escort guns.

Firepower and mobility, along with the vehicle's low height and armor plate, were important characteristics of this new weapon, which was to pass its initial tests in the 1940 French campaign.

4. As an auxiliary weapon of the infantry, the assault gun was limited to that force's attacking speed, and exposed to the fire of armor-penetrating weapons to a far greater extent than were tanks. For this reason, the armor plate had to be arranged so that it offered protection against armor-piercing ammunition up to a caliber of 3.7 cm.

5. The elevation of the assault guns should be as low as possible, and not exceed the height of a walking infantryman.

6. All the same, the interior space had to be made big enough to give the four-man crew, especially the driver and the two gunners, sufficient room to operate the 7.5 cm assault cannon.

The essential requirements were fulfilled with the design of the assault gun as a casemate armored vehicle.

The barrel of the 7.5 cm assault cannon (Stu G 7.5 cm-K) was 1307.5 mm long (= L/24); the entire weapon weighed 490 kilograms, had a traverse field of 24 degrees and an elevation field between minus ten and plus twenty degrees. At first the Panoramic Telescope 32 for Stu G 7.5 cm-K (4 x 10 degrees) was used as an aiming device. The gun fired ammunition that had been introduced for the almost identical Kampfwagenkanone IV. This was primarily the 7.5 cm shell, which carried a 5.74 kg charge up to 1000 meters at low height, and at an elevation of 20 degrees reached targets at a distance of 6200 meters (flight time 23.5 seconds). There were also two versions of 7.5 cm foglaying shells, which were used particularly to blind nests of resistance.

To fulfill tasks in the sense of Manstein's requirements as the "outstanding means of offensive tank defense", only the 7.5 cm Panzer red cannon shell was available at first.

Because of the short barrel length, the 7.5 cm assault gun was often called the "stump cannon". Early on, the vehicle's armament with a long-barreled gun was introduced. A first test model of the Armored Self-propelled Mount III with a 7.5 cm Cannon L/41 was to be introduced in May 1940.

Armored self-propelled mounts of Assault gun Battery 640, of the Infantry Regiment (mot.) "Grossdeutschland", seen in May 1940 during the western campaign. As the 16th (assault gun) Company, it formed the basis, along with Assault Gun Unit 192, of the subsequently formed "Grossdeutschland" assault gun unit.

But its penetrating power against armor plate was meager. At 100 meters it penetrated 41 mm of armor plate, and at 1000 meters 34 mm. Better prospects of successful antitank operations were provided with the introduction of the 7.5 cm Shell 38, which was available from June 1940 on. The hollow-charge shell weighed 4.5 kg, had an initial velocity of 452 meters per second, and could be utilized against armored targets at distances up to 1200 meters. Regardless of the distance, it could penetrate 40 mm of armor plate.

To attack living targets at up to 200 meters, there was the 7.5 cm KWK cartridge. It was delivered ready to fire. Its casing was destroyed in the gun barrel while being fired. A drive disc then flung some 240 lead balls out of the barrel at a scattering angle of 15 degrees. In all, the ammunition supply of the assault gun included 44 shells, most of them being explosive shells.

According to an overview of the state of the Army's equipment, nine 7.5 cm assault cannons for self-propelled mounts were installed on 5/ZW

chassis by March 1, 1940. The supply of finished assault guns rose to eleven by the end of March 1940. This figure includes the assault guns on 2/ZW chassis, which were not suitable for use in the field. The low production figures naturally decreased the speed at which the assault guns ere used as weapons, and their numbers remained far behind the original plans. According to them, every infantry division of the Army, and later also the reserve division, was to include an assault gun unit of three batteries, each with six assault guns. In May 1940 there were, including the reserve divisions, over 150 infantry and mountain rifle divisions in existence. For them, over 1700 assault guns should have been produced. On the other hand, in the first half of 1940 the Artillery Instructional Regiment in Jüterbog had set up only six assault gun batteries, though the war strength called for 445. Six assault guns belonged to one battery. Of the assault gun batteries being established, there were, at the beginning of the French campaign on May 10, 1940, three ready for ser-

The construction of a second series of self-propelled mounts for the 7.5 cm assault cannon began at Alkett in June 1940, with twelve being built. By the time winter began that year, the monthly quota was to be raised to 36 units. The picture shows the manufacture of armored hulls.

The assembly of assault guns was sometimes done along with that of Panzer III tanks. Difficulties in hull production by suppliers compelled Alkett, in June 1940, to use eight Panzer III hulls for the assault guns. The bow plates, 20 cm thick, were screwed on over the usual plates, which were 50 mm thick.

In September 1940, 24 Assault Gun III units were produced.

Installing the tracks during final assembly. At first tracks 360 mm wide were used, but they were replaced by the Kgs 61/400/120 type, 400 mm wide. Thus wear on the rubber coverings of the dual road wheels could bee minimized.

The Assault Battery (from K.St.N.445), 1939

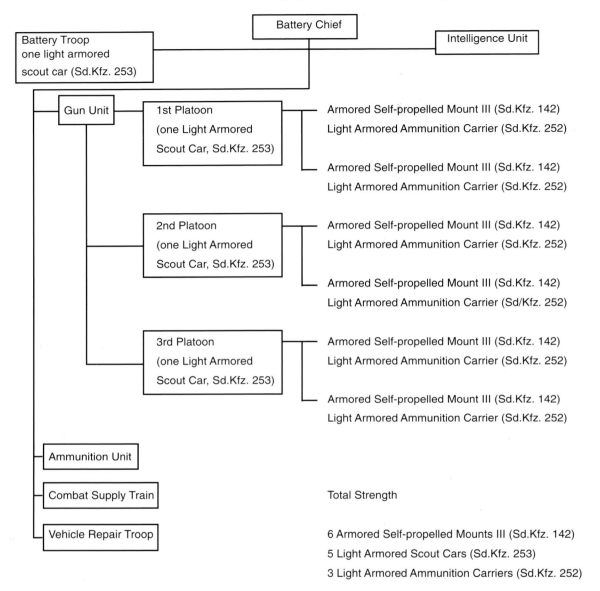

Battery Chief

Intelligence Unit

Battery Troop
one light armored
scout car (Sd.Kfz. 253)

Gun Unit

1st Platoon
(one Light Armored
Scout Car, Sd.Kfz. 253)

Armored Self-propelled Mount III (Sd.Kfz. 142)
Light Armored Ammunition Carrier (Sd.Kfz. 252)

Armored Self-propelled Mount III (Sd.Kfz. 142)
Light Armored Ammunition Carrier (Sd.Kfz. 252)

2nd Platoon
(one Light Armored
Scout Car, Sd.Kfz. 253)

Armored Self-propelled Mount III (Sd.Kfz. 142)
Light Armored Ammunition Carrier (Sd.Kfz. 252)

Armored Self-propelled Mount III (Sd.Kfz. 142)
Light Armored Ammunition Carrier (Sd/Kfz. 252)

3rd Platoon
(one Light Armored
Scout Car, Sd.Kfz. 253)

Armored Self-propelled Mount III (Sd.Kfz. 142)
Light Armored Ammunition Carrier (Sd.Kfz. 252)

Armored Self-propelled Mount III (Sd.Kfz. 142)
Light Armored Ammunition Carrier (Sd.Kfz. 252)

Ammunition Unit

Combat Supply Train

Vehicle Repair Troop

Total Strength

6 Armored Self-propelled Mounts III (Sd.Kfz. 142)
5 Light Armored Scout Cars (Sd.Kfz. 253)
3 Light Armored Ammunition Carriers (Sd.Kfz. 252)

Note: Assault Gun Battery 601, instead of the three Light Armored Ammunition Carriers with Sd.Anh. 32 trailers, had Armored Ammunition Carriers of Panzerkampfwagen I chassis (Sd/Kfz. 111)

vice (No. 640, 659 and 660). Battery 640 had already been assigned to the "Grossdeutschland" Infantry Regiment in mid-April 1940, so as to gain experience in cooperation with the infantry within that troop unit. One other, Assault Gun Battery 665, could only be sent on the march to the front on June 10, 1940, It saw service in the Vosges, where it had to prove itself by penetrating a line of French bunkers. In all, 24 assault guns were in service in France; four of them were lost.

The experience gained in the French campaign resulted in numerous service reports, soon followed by initial printed orders for the further instruction of the troops. Assembled by the officers of the VI. Instructional Unit and published in the summer of 1940 in the brochure "Die Sturmbatterie—Einsatz- und Ausbildungserfahrungen", these tactical and technical experiences made important contributions to the development of the assault gun weapon. At this point it was characterized as follows: "The Pak battery (Sfl.), (Assault batteries), have 7.5 cm cannons on the chassis of Panzerkampfwagen III. Thanks to its heavy armor and off-road capability, the assault gun, as compared with the guns of the division's artillery, are in a position to follow their own infantry or the Panzer troops everywhere. The main tasks of the assault gun are: Fighting down heavy enemy infantry and antitank weapons that cannot be destroyed by our own heavy infantry weapons. The assault battery usually fights in battery or platoon form, as well as to provide protection after combat tasks that it is given by the infantry or the Panzer troops. It is fired mainly in direct fire." Later in the same document it is stated: "In exceptional cases, the assault battery can also be used as division artillery. Its use then is based on the principles for the I.F.H. battery. . . ." Similar statements typify the contents of the H.Dv.200/2 m "Vorläufige Ausbildungsanleitung für die Sturmbatterie" (Temporary Training Directions for the Assault Battery), which likewise was published in 1940.

The insistence on the use of assault batteries as division artillery is interesting. It was related to the concept that the artillery-s typical type of shooting was indirect fire. For this, the 7.5 cm cannons of the assault guns were, of course, suited to the technical prerequisites, but in practice they were seldom used. As a rule, assault guns attacked their assigned targets with direct fire. During the course of the war, this was true to such an extent that for assault artillerymen who were ordered to Germany for training in 1943, indirect firing was regarded as something unusual.

Another point from the training and action principles of 1940 is worthy of special note, namely the fact that infantry and armored troops were supposed to be supported equally by the assault guns. From the beginning, assault guns were conceived and developed as infantry support weapons, and primarily because they could not get along completely without armored vehicles of their own. It seems contradictory to assign such vehicles for fire support to the Panzer troops, whose tanks were ideal combinations of firepower, mobility and armor, and thus necessarily to take them away from the infantry. The increasing numbers of the assault gun weapon and further combat experience led in the ensuing war years to a clear exposition of purposeful combat principles for the assault guns.

In expectation of enemy air attacks, the distances between the vehicles were doubled on the march. In attacking and observing the enemy as well, there was a loosening of the marching order or the arrangement of the battery.

In the assault gun batteries, the usual marching order was in the battery column. The battery troop went first, followed by the three gun platoons, and ammunition unit, the supply trains (I and II), and the vehicle repair unit.

Assault Artillery—
the Infantry's Armored Fist

The first independent unit of the assault artillery, Sturmgeschützabteilung 184, was established in the late summer of 1940. Its three batteries were formed in accordance with War Strength Directive 446 (assault battery at unit strength). Every battery consisted of three platoons, each with two Armored Self-propelled Mount III (Sd.Kfz. 142) and two Light Armored Ammunition Carriers (Sd.Kfz. 252). The Light Armored Scout Cars (Sd.Kfz. 253) planned for in War Strength Directive 445 of November 1, 1939, for the three platoon leaders and battery chiefs, were partially eliminated. In 1941 the battery leaders were assigned an Armored Self-propelled Mount III (Sd.Kfz. 142), which were available in sufficient numbers by then. The designation of the vehicle was also made uniform, even though the names of 7.5 cm Assault Cannon on Self-Propelled Mount, Assault Gun 7.5 cm Cannon, or Armored Self-propelled Mount were still used for a long time in writing as well as orally. In the already cited brochure "Die Sturmbatterie—Einsatz und Ausbildungserfahrungen" of the summer of 1940, it is even referred to as Pak (Sfl.), a concept from this vehicle's developmental period. From now on the designation of Sturmgeschütz was the basis. By the spring of 1941, Assault Gun Unit 184 had been followed by three more units, 185, 190 and 191. Three of these saw service in the Balkans in the course of that campaign. With the given geographical conditions, essentially characterized by steep, jutting rocky mountains, deep valleys and few roads, most of them winding, narrow trails, the use of the assault gun turned out to be very complicated. If the vehicles came under fire, they could not turn. Driving them backward, what with the lack of a view to the rear, had to be done with great caution, so as not to lose the valuable vehicle. They could fire only in the direction in which they were going, and then only in a narrow sector (24 degrees) allowed by the cannon's field of traverse. Turning the gun to the side was often impossible because of the particular nature of the roads. In addition, the goals to be attacked for the infantry's sake (bunkers, machine-gun nests) were often at higher elevations and could not be reached by the fire of the assault guns.

An armored self-propelled mount for the 7.5 cm assault gun of Assault Gun Unit 191, seen in Bulgaria in the spring of 1941.

On the left track apron of this assault gun, the symbol for Assault Gun Unit 191, a bison, can be seen.

Assault Gun Unit 191 on the march to the Greek border. The attack of the German troops on Yugoslavia and Greece began on April 6, 1941, at 6:15 AM. Heavy fighting on the Metaxas Line ensued, into which this unit was drawn.

A pause in the march of Assault Gun Unit 191. The crews remained on the vehicles and took their meals there. This picture was taken early in April 1941.

A look through the rear door of the Light Armored Scout Cat, Sd.Kfz. 253, of a platoon leader of the I./Assault Gun Unit 191. Note the details of the door latch.

The armored support vehicles of the assault gun units began to show shortcomings. The armor plate of the Armored Scout Car (Sd,Kfz. 253) turned out to be too thin. Wherever the assault guns appeared, they naturally drew the fire of the enemy's antitank weapons and artillery to themselves, under which the scout cars had to suffer all the more. Much the same was true of the Armored Ammunition Carrier (Sd.Kfz. 252) and Special Trailer (Sd.Anh. 32d). Crucial losses were the result.

Despite many problems, it was seen in the course of the Balkan campaign that the assault guns could provide the hard-fighting infantry with valuable support. This can also be seen in inclusive reports, such as that of the 72nd Infantry Division. This division was supported by the first battery of Assault Gun Unit 191 in the combat at the Greek Metaxas Line in the area of the bunkered Mount Ochiron early in April 1941.

In the entire year of 1940, only 184 assault guns had been delivered. Their numbers rose in 1941 to 548. In January 1941 alone, the number of delivered assault gun vehicles was 44 (36 planned). In February 1941 31 7.5 cm assault cannons were mounted and delivered. The significance that the assault gun had gained by then makes the fact clear that for the B series built as of June 1940, Panzerkampfwagen III chassis (Type 5/ZW) had to be prepared in order to reached the planned numbers in assault gun production. Technical improvements were constantly being added. The C version of the assault gun, delivered as of March 1941, was fitted with a telescopic sight for self-propelled mounts, which was installed above the roof of the armored body. With that, the previously used optical shaft on the left side of the body, which had been sensitive to shot damage, was eliminated.

Between March 1940 and June 1941, 285 chassis had been produced for the Armored Scout Car Sd.Kfz. 253. The idea of sending the platoon leaders to the combat batteries in special command cars, only lightly armored, along with the assault guns did not work out well. From the viewpoint of artillery, it was based on the concept of having a mobile observation and command vehicle separate from the guns.

To be able to supply the armored self-propelled mounts of the 7.5 cm assault cannon with ammunition on the battlefield, 413 lightly armored ammunition carriers (Sd.Kfz. 252) were built and assigned to the assault gun batteries between June 1940 and September 1941. These vehicles were often used by the medical staffs to rescue wounded men.

Assault Gun Battery 660, which had been established at Zinna in April 1940, received Ammunition Tank I with Special Trailer 32 instead of Sd.Kfz. 252. These soon proved to be unusable and were eliminated.

With assault gun production running at capacity and the technical improvement of the vehicles, important material requirements for building up the assault gun weapon were provided. On June 1, 1941 they consisted of the five independent batteries already mentioned (659, 660, 665, 666 and 667). The number of units had grown to twelve (No. 184, 185, 189, 190, 191, 192, 197, 201, 203, 210, 226 and 243). At this time, Assault Gun Units 177, 202 and 244 were being formed, and others were planned. In addition there were the "Grossdeutschland" Assault Gun Company and a battery of six guns in the motorized "Leibstandarte Adolf Hitler" Infantry Regiment of the Waffen-SS.

The personnel of the assault gun units consisted until 1943 exclusively of volunteers, who brought with them, among other things, their superb training in firing techniques, some of it frontline experience. In many cases the drivers of the assault guns were also volunteers from the replacement and training units of the Panzer troops.

Noteworthy and surely not without significance for the further development of the assault gun weapon were the conferences of military leaders. For example, one such took place on June 6, 1941,

between the Chief of the Army General Staff, Generaloberst Franz Halder, and the Commander of Panzer Group 2, Generaloberst Heinz Guderian. The tasks of the Panzer groups in the forthcoming campaign against Soviet Russia were discussed. Halder noted in his diary: "The main task of the Panzer groups is not in launching attacks but in depth. For this task, forces must be maintained. Therefore, when launching an attack, support by infantry forces is to be utilized." The Panzer divisions were not to be worn out in attacks on an enemy ready to defend himself. Infantry divisions were to be foreseen for this purpose. It was their task to break holes in the enemy defensive line through which the tanks could then advance into the hinterlands. For this difficult task the infantry needed the support of armored vehicles, but these were not to be taken from the Panzer divisions, as the assault guns were available and nearly ideal for the task. From this standpoint, Guderian, who was appointed Inspector General of the Panzer Troops barely two years later, understood the purposes of the assault artillery. It is obvious that their further development had to take place at the expense of the Panzer troops.

In the French campaign of May and June 1940, the assault gun units had to depend on the repair units of the Panzer troops to have serious damage to the self-propelled mounts repaired. Only in 1941 were they, and a little later the towing staffs, assigned the 18-ton Heavy Towing Tractor Sd.Kfz. 9) as well as the 22-ton Low Loader Trailer for Tanks (Sd.Ah. 166), shown third from right.

As of September 1941, the armored self-propelled mount for the 7.5 cm assault gun was delivered in Type E form. In all, 275 units of it were built.

In order to be able to carry additional radio equipment for optional use as a platoon leader's or battery chief's gun, the extension built only on the left side until then was also added to the right side. In place of the radio equipment, six more rounds of ammunition could be carried.

An assault gun battery on the way to its readiness area. They were to support the infantry units at focal points in penetrating the enemy defenses. Into the holes thus broken, the Panzer divisions could then move to strike the enemy's rear. This picture was taken on the eastern front in July 1941.

Arrived at the readiness area, the combat unit of the assault gun battery prepared for the attack. The situation and terrain could make it necessary to have the supply staff located there as well. Their vehicles stayed back when the assault guns advanced further.

An assault gun battery in its initial position. It was taken up just before the attack began in order to surprise the enemy. The use of closed batteries was to be striven for. In H.Dv. 200/2 m "Die Sturmgeschützbatterie" it was stated: "Assault guns are a focal-point weapon. Their effective possibilities are fully utilized only when a division of the battery into platoons or single guns is avoided."

Facing New Tasks

On June 22, 1941, the campaign in the east began. Assault guns were detailed to the infantry divisions that fought at focal points in the battles. The increasing losses of men and materials told of the extraordinary hardness of the combat. By the year's end, 95 assault guns had to be written off as total losses (on hand early in June 1941: 377), The number of vehicles that were out of action for long periods on account of enemy action or wear can be estimated as being much greater. The active strength of the assault gun units hit ac dramatic low point for the first time. In a report on the experiences gained in the eastern campaign it was stated: "The eastern campaign brings decisive changes for weapon and ammunition production. It was shown in the course of the war in the east that the consumption of materials rose far above the previously known extent. The wear on weapons is very great. . . ."

The assault artillery's main task was the support of the infantry in attacks. To be sure, since the first day of the eastern campaign, the assault artillerymen were applied to a great extent to defend against Russian tank attacks—a task that was considered when this weapon was originated, but on which emphasis was not placed and for which the important technical prerequisites (long-barreled guns) were lacking. The reasons for the utilization of assault guns in antitank defense were manifold. The Russian armored troops, very strong in numbers and partially equipped with the most modern tanks, encountered the antitank forces of the German Army, clearly inferior in terms of weapon technology. They presented themselves in the 14th (antitank) companies of the infantry regiments with nine 3.7 cm antitank guns, and in the antitank units of the infantry divisions with, at best, a mixed assortment of, in all, 36 3.7 and 5 cm antitank guns. At a range of 500 meters, these guns, with a striking angle of 60 degrees, could penetrate armor plate 29 to 59 mm thick. This was too little against the T-34 tank. The artillery regiments in the infantry divisions, which could be called in for antitank defense, should also be noted. Their being equipped almost completely with howitzers made their use for this purpose considerably more difficult.

The assault guns units were sent to the east by rail in the early summer of 1941. The crews remained with their guns during the trip.

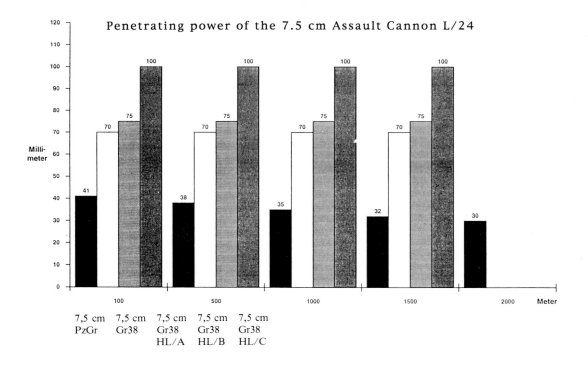

Penetrating power of the 7.5 cm Assault Cannon L/24

	7,5 cm PzGr	7,5 cm Gr38	7,5 cm Gr38 HL/A	7,5 cm Gr38 HL/B	7,5 cm Gr38 HL/C

Assault guns, made with cooperation with the infantry in mind, themselves armored and also mobile, became rescuers in times of need for the infantry divisions more and more often, what with the weaknesses of antitank defenses. Their weapons were at first of only limited use against the Russian T-34 and KW tanks which appeared in greater and greater numbers. For armor-piercing ammunition, the 7.5 cm Panzergranatpatrone rot, introduced even before the war, was available (38 mm armor penetration at 500 meters range, with a striking angle of 60 degrees), as was the 7.5 cm Granatpatrone 38 as of June 1940. The latter could penetrate 40 mm regardless of the distance. These shells provided conditions that allowed the assault gun crews to fight Russian tanks at "suicidal close-combat distances". At a range of 100 meters, some penetration of the front and side armor of the T-34 could be achieved. The assault guns had worse prospects of success when they met the heavy KW tank. At the end of 1941, improved hollow-charge ammunition was introduced. At first it was the 7.5 cm Granatpatrone 38 HL/A, then the HL/B, which could penetrate 70 and 75 mm of armor respectively. Now antitank warfare could be carried out more successfully at ranges under 1500

meters. Yet four to six shots were still necessary to score one hit.

The existing technical inferiority, and even more the numerical inferiority had to be made up for by better leadership and fire control, as well as by outstanding firing training for the assault gun teams. Peculiarities of the assault gun weapon, including, as already mentioned, the close cooperation with the infantry, were surely an advantage as well. The high mobility of the assault guns allowed focal points to be built up quickly, and their low bodies made them hard to spot and fight against; and finally, the technical layout as a casemate tank made it possible in many cases to await attacking tanks in an advantageous firing position.

Something basic to the role of the assault gun as an army weapon and to its future development was said on November 29, 1941, in the course of a discussion on the subject "Tank production and antitank defense", which was attended by Reich Chancellor Adolf Hitler and high-ranking representatives of the Wehrmacht, the Ministry for Weapons and Ammunition, and the industry. On page 5 of the protocol it is stated: "The construction of assault guns is proceeding independently of that. The goal must be giving the motorized divisions the Panzer spearhead by assigning armored vehicles to them and thus replacing the assault guns, used for this purpose at this time. The assault vehicles belong to the infantry, which must absolutely accompany them. The assault vehicles must contain a gun that has high effectiveness against tanks. The assault guns must receive more ammunition. The speed of the assault gun can be decreased if it is used only as an infantry escort gun and is released from tasks that belong to the tank weapon. For its use with infantry divisions, a speed of 12 kilometers per hour is sufficient."

One fact was proved on the eastern front in the first war winter of 1941-42: when it was not possible to get the upper hand over the constant threat of tank attacks, the infantry divisions were no longer in a position to take offensive action. The assault gun took on a vital role in antitank defense. In an evaluation of December 13, 1941 is was also stated: "Assault guns are the absolutely necessary means of upholding and increasing the attacking power of the infantry; every infantry division should, as a final goal, receive three assault gun batteries." An idea from prewar days. The realization of such ideas foundered on the limited manufacturing potential of the German armament industry. There were also increasing

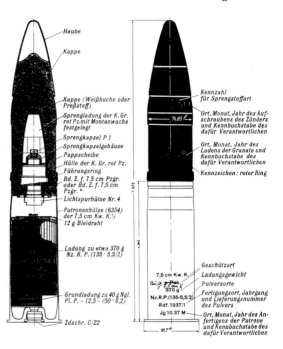

Ammunition of the 7.5 cm tank gun
7.5 cm antitank shell for tank gun

Front view of an armored self-propelled mount for assault gun, with a 7.5 cm cannon, Type D. As of May 1941, 150 of these vehicles were delivered. Several were used in North Africa by Special Unit z.b.V. 287.

Assault guns and motorized infantry on the march in the east in the summer of 1941. "The beaten enemy is to be pursued without regard for man or material," the instructions said. In many cases the chiefs of the assault gun batteries were the leaders of the pursuit troops.

The ruthless action of the assault gun weapon in the eastern theater of war resulted in increased vehicle breakdowns under the prevailing geographical and climatic conditions.

The Assault Battery (from K.St.N.446), 1941

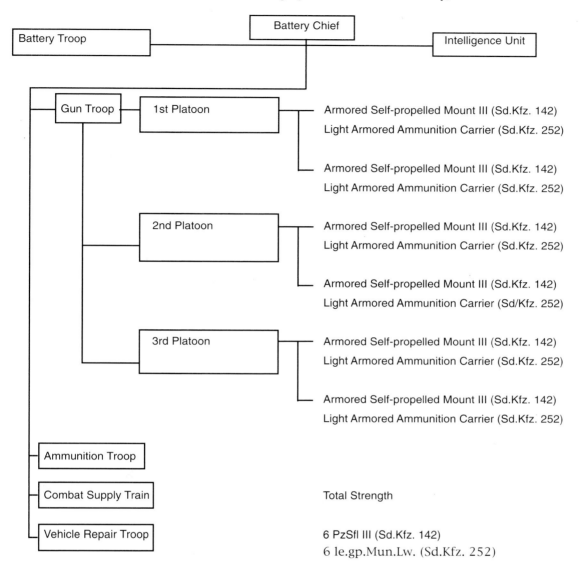

Total Strength

6 PzSfl III (Sd.Kfz. 142)
6 le.gp.Mun.Lw. (Sd.Kfz. 252)

Note: The batteries established according to War Strength Document 445 of 1939 still had five Light Armored Scout Cars (Sd.Kfz. 253). They were retained, also after August 9, 1941, when according to Befehl Org. Abt./OKH, one Armored Self-Propelled Mount III (Sd.Kfz. 142) was assigned to the battery troop for the battery chief. The Light Armored Ammunition Carriers (Sd.Kfz. 252) were partially replaced by light trucks.

losses of material. In the first three months of 1942, 63 assault guns were lost. That equaled two thirds of the entire losses in the year of 1941.

The first considerations for developing a long-barreled 7.5 cm tank gun as the primary weapon for the armored self-propelled mount were made in 1938. At the beginning of the year, the Krupp AG firm displayed a test model. In March 1941, for the first time, a long heavy antitank gun which was the official designation of the 7.5 cm Cannon L/41, was installed in the Armored Self-propelled Mount III and displayed to Adolf Hitler.

Quite obviously under the impression of the appearance of modern tanks on the eastern front, the people at the Army Weapons Office came to the conclusion that the gun then being tested had too meager performance reserves, and in Novem-

For the 7.5 cm assault gun, the best shot ranges were stated as 400 to 1000 meters. At great distances three shots were supposed to be enough to achieve the effect of one direct hit; at medium distances (1000 meters) it took five to eight shots. For combat against the heavily armored Russian tanks, the use of ammunition was considerably higher.

In the summer of 1941, the Russian T-34 tanks were an unpleasant surprise for the German troops. By June 1941, 1119 of these modern tanks had been produced. With the short 7.5 cm assault gun, only a hindering effect could be achieved against the front armor of this tank. Greater chances of successfully combating them came with the use of hollow-charge shells at the sides and rear. With the 7.5 cm antitank shell, armor could be pierced at ranges under 100 meters.

ber 1941 they had the work on it halted. In the meantime, the development of a long assault gun had been urged; it was originally designated 7.5 cm Cannon 44 L/46. Its relationship to the weapon that later became known as the 7.5 cm Panzerjägerkanone 40 L/46 was not hard to recognize. On March 16, 1942 the 7.5 cm Cannon 44 L/46 was renamed the 7.5 cm Assault Gun 40 L/43. It was not able to take the same shell casing that was used in the 7.5 cm Tank Gun 40 L/43 of the Panzer IV tank. Both weapons were fitted with the bottle-shaped muzzle brake. The first three guns of this type were installed in assault gun mounts in February 1942. They showed several shortcomings. After the first shots were fired, casings often got stuck, which limited their combat usefulness. Sometimes the casings had to be pushed out from the front after every shot.

Up to this point, assault guns had been produced with the short 7.5 cm cannon. These were Types B, C and D, of which 520 had been completed in all. The Type E was delivered as of September 1941. 275 of these were built.

In March 1942, the first series-produced assault guns with the new long-barreled cannons rolled off the assembly line. Some of them were created by rearming Type E Assault Gun III vehicles. The official designation was "Armored Self-propelled Mount for Assault Gun 7.5 cm Cannon, Type F". By September 1942, 364 units had been completed. At the front, these vehicles were used at focal points in the southern part of the eastern front, with Army Groups A and B. 182 Type F assault guns reached the troops with additional 30 mm armor plate.

The 7.5 cm Assault Cannon 40 L/43 fired armor-piercing Panzergranate 39 shells with an initial velocity of 740 m/sec, and could penetrate a 91 mm armor plate at a range of 500 meters. The Panzergranate 40 shell could even penetrate 108 mm. The latter type was used only rarely. The Panzergranate 39 was sufficient.

According to Russian estimates, assault guns proved to be "effective means of supporting the infantry". Soon the Red Army was using similar combat vehicles in large numbers.

In defensive action, assault guns were not supposed to be used deliberately as bunkers on the main battlefield and draw the fire of the enemy's heavy weapons to themselves. The armored self-propelled mounts were rather to be used to support the infantry's counterattacks.

Assault guns accompany the advance of the infantry in the east in 1941. As an offensive weapon to support the infantry, the assault guns had given fine service in all theaters of war up to then. Now, though, antitank defense became more vitally important. For that the short 7.5 cm assault gun had only limited success.

An assault gun in a small Russian city, seen in July 1941. This was a Type B vehicle, the production of which had already ended early in 1941.

In the course of the heavy fighting in the east, the combat strengths of the assault gun units decreased sharply. Assault Gun Unit 190 lost eight self-propelled mounts and five ammunition carriers and command cars in three months of the eastern campaign in 1941. In the same period, 15716 rounds of 7.5 cm ammunition (equaling 94.3 tons, without packing containers) were fired.

The armor of the assault guns (Type B is shown here) was 50 mm thick in the front (at 69, 80 and 75 degrees of inclination). Thus within 500 meters it offered neither protection from the shells of the Russian 45 mm antitank cannon, 1937 model (penetrating power 57 mm or armor plate at 500 meters), nor from the 76 mm tank guns and field cannons of the Red Army. These facts resulted in the strengthening of the front armor to 80 mm in 1942.

Muffled in a thick cloud of dust, an armored self-propelled mount (Type B) passes a Büssing-NAG truck on the unpaved summer lane of a paved road.

The comparatively slow marching speed of the infantry burdened the engines of the assault guns. In addition, their fuel consumption was comparatively high. Therefore the assault gun units were stopped and sent forward spasmodically to catch up to the infantry, as shown in this picture, taken on the eastern front in September 1941.

Full speed ahead! The top speed of the assault gun was stated at 40 kph. Its range, depending on weather conditions, was between 95 and 165 kph. The high marching performance in the east resulted in only a fraction of the vehicles of an assault gun unit being capable of action after three to four weeks of service.

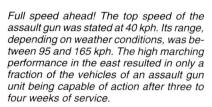

No changes were made to the hollow-charge ammunition still in use. With this weapon and this ammunition, it was possible to fight the Russian T-34 tanks at ranges from 500 meters in front and 1000 meters to the side with good prospects of success. In a report of a special command of the Army Weapons Office of May 20, 1942, it was said of the action of six assault guns with the long barrels that were being used by Assault Gun Unit 190 at that time: "Weapon effect the same as with Panzerkampfwagen IV with Kampfwagenkanone 40. . . . By each of the six new assault guns, three Russian tanks were shot down." The effects of shots on thirteen tanks, including three T-34s, were tested on the battlefield. Most of them showed complete penetration, especially on the front of the turret and the sides of the hull. Many vehicles were completely burned out, and damage to the weapons and the running gear could also be seen. The greatest distance at which tanks could be fired on with success, was 1200 meters. Thus the assault gun had become one of the most effective antitank weapons of the Army.

The important technical changes were not considered in the Artillery Memorandum No. 34, Guidelines for the Use of Assault Gun Units, of April 27, 1942. These were to be valid until a new edition of Army Service Instructions 200/2m, Temporary Training Instructions for the Assault Battery, would be issued. In spite of that, the nature and tasks of the assault gun in Memorandum No. 34 were stated considerably more clearly as had been the case, for example, in the summer of 1940. It was stated: "The assault gun—a 7.5 cm cannon in an armored self-propelled mount—is an offensive weapon...Off-road capability, armor protection, mobility and constant readiness to fire allow the assault gun to accompany the infantry's attacks and give it a constant artillery support at an effective combat range. Assault guns lend the attack might and speed and give it a strong support in terms of morale." The use of the assault guns as division artillery or their cooperation with tanks was expressly forbidden, because they did not correspond to the tactical and technical possibilities of the weapon.

A typical picture of assault guns in service. In an indefinite combat situation, the infantry seeks cover behind the armored self-propelled mounts. Their close cooperation was a prerequisite for the success of them both.

The necessity of drawing assault guns more strongly into antitank defense still was given insufficient consideration. Under Point 3 it was stated only that, "In combat against armored vehicles, assault guns can fight light and medium tanks successfully." The structure of the assault gun unit and the assault gun battery portrayed in Memorandum No. 34 still corresponded to that of War Strength Directive 446 of April 18 and November 1, 1941. According to it, the assault gun unit consisted of the unit staff with the staff battery and three assault gun batteries of seven guns each (three platoons of two guns each and one gun for the battery leader). Otherwise the structure resembled that of a light motorized artillery unit.

Basically, the assault gun unit was to be subordinate to the infantry commander whose troops it was to support. The division of the assault gun units into smaller units (platoons, individual guns) limited the firepower and made enemy defense easier. The action of individual platoons was therefore to be limited to exceptional cases. All too often, this last clause remained a pious wish in practice. Individual assault guns were often were often brought into position behind infantry position as "corset stays of morale", and often enough they thus became easy spoils for enemy tanks or artillery. The dwindling quality of the infantry, at this point, already made cooperation with assault guns more difficult.

A readiness position for infantry and assault guns on a slope, out of sight of the enemy.

The Armored Ammunition Carriers (Sd.Kfz. 252) of the combat unit followed the assault guns as soon as the combat situation and need for ammunition made it necessary. Eastern front, summer 1941.

In the assault gun batteries, each self-propelled mount was assigned a combat supply of 300 rounds. Of them, only a small number (44 antitank shells) were actually carried with the gun. Most were carried by the Armored Ammunition Carriers (Sd.Kfz. 252) and their Ammunition Trailers (Sd.Ah. 32). Some of these rigs were already replaced in the summer of 1942 by the Light Armored Ammunition carrier (Sd. Kfz. 250/6) and the Sd.Ah. 32/1 trailer. In the supply train, the ammunition troop also had four medium off-road-capable trucks.

Assuring the timely supplying of the assault guns with ammunition was the responsibility of the battery chief. As a rule, the armored ammunition carriers pulled up next to the assault gun mounts to transfer the antitank shells.

7.5 cm Explosive Shell 34 ammunition is being transferred from the special trailer to the assault gun. Every shell weighed 7.35 kilograms, and 49 rounds were carried. The one-axle trailer for ammunition (Sd.Ah. 32) weighed 780 kg when carrying a 390-kg payload. Its dimensions were 2160 x 1700 x 1100 mm.

Thus in an experience report of the XXXXIV. Army Corps on March 31, 1942 it was stated that the assault guns had proved themselves again and again in supporting German attacks and counterattacks. It was stated further: "Cooperation with the infantry often left something to wish for, on account of the almost complete lack of experienced group, platoon and company leaders."

The bad experiences of the 1941-42 winter campaign inspired the Army General Staff to prepare for the expansion of the Army troops at the right time. In a letter from Operations Department (III) of September 3, 1942 (No. 34 149/42gKdos) it was stated: "The main problem on the eastern front continues to be defense against enemy tank attacks." To do this, mobile units that were not attached to the division units (assault gun or mountain rifle units), above all, had to be made more numerous and more ready for action.

Additional assault gun units were established. The number of assault guns continued to increase despite considerable losses. In June 1942, nineteen assault gun units and one battery were in action on the eastern front. On April 1, 1942 there were 623 assault guns on hand, and the number rose to 1422 by June 30, 1943. A large number of them were produced in 1942, for which the stated total is 788 (93 of which still had the short gun barrel). The increase was achieved despite the loss of 330 assault guns. At the end of 1942, the monthly production reached 120 vehicles, and in the new year it jumped upward again. In April 1943 alone, 262 assault guns were delivered to the troops. Among them were thirty-four 10.5 cm assault howitzers, production of which had begun with five units that January.

On November 1, 1942 a new war strength standard (K.St.N. 446a) came into effect. According to it, the assault gun battery was divided into three platoons with three guns each. There was also a gun for the battery leader. With the vehicles for the other two batteries and that of the unit commander, the assault gun unit now had the use of 31 assault guns. This was a considerable increase in fighting power, if one compares this number to the figures in the old war strength standard of November 1941. Then a unit had only 22 assault guns. The old rule remained valid for a transitional period and only went out of power at the beginning of 1943.

The numerical strengthening of the assault gun units made it possible, in the practice of dividing batteries among individual regiments of a division (and beyond that!), still to be able to provide effective firepower. It is also worth noting that the assault gun units used their vehicles and weapons more and more to the point of breakdowns. After three to four weeks of unbroken front action, the number of usable guns sank to 20% to 30% of the specified strength. Here is an example: According to the daily report of General Command, XXX. Army Corps, in the second half of June 1942 Assault Gun Unit 249, with a specified strength of 21 guns, only had eleven guns ready for action on two days. On June 13, eight guns were ready for action, and the absolute nadir was reached on June 16: Not a single assault gun was ready for use, five were undergoing short-term and ten long-term repairs. In many cases it was damage to the powerplant that resulted in breakdowns. Still in all, the numerical strengthening of the units enabled them to keep their actual strength at a somewhat higher level.

The troop emblem of Assault Gun Unit 192.

An assault gun battery on the march. As a rule, the battery chief's gun traveled with the first platoon, which also led the unit in case of unified action with an infantry unit. In combat, its firing activity was to be limited to a minimum. Command tasks were more urgent.

In the spring of 1942, the first assault guns with long-barreled cannons were assigned to the assault gun units, particularly those to be used at the focal points of the 1942 summer campaign. Thus Assault Gun unit 190 of the 11th Army received the first six long-barreled guns for Operation "Trappenjagd" (May 8-18, 1942) on the peninsula of Kertsch. This picture shows armored self-propelled mounts with 7.5 cm assault guns, Type F, in Assault Gun Unit 191 in the summer of 1942.

In numerous assault gun units, the guns with the short barrels were kept in service until the spring of 1943. When fighting with tanks, their crews had to rely on the hollow-charge ammunition (7.5 cm Granatpatrone 38/Hl/B) that had meanwhile been improved.

The core divisions of the SS troops were already given their own assault gun batteries, which were later expanded into units, in 1940-41. This picture shows an armored self-propelled mount with a 7.5 cm cannon, Type B, used by the motorized SS division "Das Reich". The low body of the assault gun is easy to see.

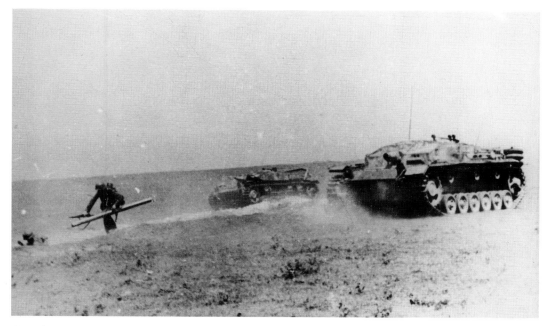

Assault guns played an important role in the combat for the Russian defensive lines and positions, such as the penetration of the "Stalin Line" in 1941 and the fighting for the fortress of Sevastopol in 1942. The picture shows engineers being given fire cover while bringing in a lengthened explosive charge.

A rifle group is being seated on an armored self-propelled mount for the 7.5 cm assault cannon in Assault Gun Unit 245. It was established in Defensive Zone III on June 13, 1941, and was wiped out at Stalingrad in January 1943. In April of that year it was formed anew at Jüterbog.

Nr. 6

Deine
astrierte"
hast,
n Solda-
ennst, die
es Dir.

Berlin, 10. Febr. 1942

10
Pf.

18. Jahrg. Nr. 6

zuzüglich ortsüblicher
Zustell-Gebühr

This picture, showing an assault gun of the 1st Company of Assault Gun Unit 192, decorated the front cover of Number 6 of the periodical "Deutsche Illustrierte" of February 10, 1942.

The "Totenkopf" Unit (No. 192) had been formed at Jüterbog on November 25, 1940. This picture was taken on the eastern front in the summer of 1941. The assault gun has been hit on the left side by an artillery shell.

After heavy and costly fighting in the summer of 1941 as well as in the fall and winter of that year, Assault Gun Unit 192 and the former 16th (Assault Gun) Company of the motorized "Grossdeutschland" Infantry Regiment were made into the assault gun unit of that division. Thus the familiar troop emblem of the "Death's-head" unit disappeared. Here it can be seen again on a vehicle of the 3./Assault Gun Unit 192.

An Armored Self-propelled Mount for Assault Gun 7.5 cm Cannon, Type B, on the eastern front in the winter of 1941-42. For the first time, the active strengths of the assault gun units declined dramatically.

This assault gun with the striking name of "Luchs" (Lynx) was left behind by the German troops in the hard and changeable fighting before Moscow. Often there was a lack of fuel, and the extremely low temperatures also caused technical failures that the overworked repair units could not fix.

The lack of supplies of spare parts from Germany very often resulted in valuable vehicles like this armored scout car (Sd.Kfz. 253) and assault gun had to remain in the hinterlands for a long time, as here in February 1943.

To strengthen the breaking front in the east, the few heavy weapons were brought together in battle groups for the first time in the winter of 1941-42. This picture shows a Panzer 38(t) tank, a captured 740(r) tank (the Russian T-26 C) and an Armored Self-propelled Mount for Assault Gun 7.5 cm Cannon, Type E.

The infantry, equipped with skis, laboriously prepares for a counterattack. The small battle group is to be supported by an assault gun and a Panzer II tank.

December 1941, northern sector of the eastern front: an assault gun rolls through the village of Gobunki.

Towing a Type E assault gun out of a river in the Leningrad (St. Petersburg) area early in the nineties. The self-propelled mount had been lost in battle in 1941.

After the assault gun had been towed out of the water, it was temporarily cleaned up. The well-preserved gray paint came into view as a result.

Transported back to Germany, the assault gun can now be seen in the Vehicle and Technology Museum in Bad Oeynhausen.

Two Type D assault guns in the streets of a Russian city; in front of them is a destroyed Russian T-27 tankette.

Removal of the 7.5 cm assault gun. The self-propelled mount looks comparatively small in front of the 18-ton towing tractor (Sd.Kfz. 9) with makeshift crane.

The barrel and breech of the 7.5 cm assault gun. The barrel was 1766.5 mm (= L/24) long and weighed 287.5 kilograms in this state. In slightly changed form, it was used by the Panzer troops as the 7.5 cm Cannon (sfl) in the SPW (Sd.Kfz. 251/9) and the Observation Tank (Sd.Kfz. 233).

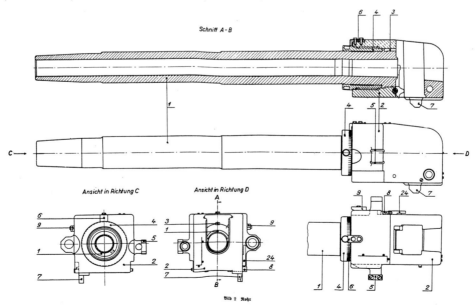

In December 1941, a prominent spokesman for the Army had spoken in favor of assault gun production: "For them too, the numbers in our plans are fully insufficient. With them we simply cannot wage war. . . ." For the operations on the south wing of the eastern front in the early summer of 1942, the self-propelled mounts, still coming from factories and repair shops in small numbers, were assigned to the assault gun units concentrated at focal points, to allow further advances into the vast extent of Russia.

In pursuit of the beaten enemy, the infantry was to be loaded onto assault guns, so as to increase the marching speed (see H.Dv. 200/2m "The Assault Gun Battery" of September 7, 1942).

The too meager production of Type E assault guns, which were slowly equipped with the new 7.5 cm Assault Cannon 40 L/43 as of February 1942, resulted in the fortunes of the assault gun units in the summer of 1942 still depending on the older-series vehicles.

Heat, dust and powder smoke were part of everyday life for the assault artillerymen on the eastern front.

An assault gun before the fortress of Sevastopol, June 1942.
For Operation "Störfang", the reduction of the fortress, Assault
Gun Units 190, 197 and 249 were on hand with the 11th Army.
Most of their assault guns had short barrels. Only Assault Gun
Unit 249 is known to have received six new long-barreled guns
in March 1942.

An assault gun is seen crossing an antitank ditch.

Advancing across the Don Valley toward Stalingrad in June 1942. The assault gun unit of the "Grossdeutschland" division and Units 177, 190, 191, 203, 243, 244 and 245 were assigned to the operations on the south wing of the eastern front. Four units (177, 243, 244 and 245) were later wiped out in the pocket at Stalingrad.

Combat in the city area of Stalingrad. Note that the Type B assault gun at this point still had the drive wheels for the 380 mm tracks.

The assault gun unit of the "Grossdeutschland" Infantry Division had received 21 new long-barreled guns at the beginning of the German summer offensive of 1942.

The inclusion of a complete assault gun unit in the "Grossdeutschland" Infantry Division (motorized) gave this elite division much greater fighting power. Strengthening all infantry divisions in this way, as ordered in December 1941, remained only a dream.

The main task of the assault gun units until 1942-43 was to work together with other service arms to support the infantry's penetration of enemy positions and advance into the depths of the battlefield.

An assault gun of Assault Gun Unit 243 carries infantrymen through the streets of Stalingrad, photographed in the autumn of 1942.

Two light armored ammunition carriers (Sd.Kfz. 252) on the steppes of Russia. Production of these halftracks, which weighed almost six tons, had already been halted in September 1941. One could still see an occasional one in action until 1943-44.

Assault guns in action against Russian tanks. Aside from the new assault cannon with L/43 or L/48 barrel length, the Type F and F/8 armored self-propelled mounts for 7.5 cm assault guns differed from the Type E version in just a few external details.

Almost simultaneously with the introduction of long-barreled assault guns, the renown of numerous leaders and sub-leaders, who rose from the ranks of the assault artillery and received high military decorations, grew quickly. They included Major von Malachowski, Hauptmann Franz and Oberwachtmeister Primozic, all of whom proved themselves in battle against Russian tanks. Here Oberwachtmeister Primozic is seen with his crew at Jüterbog. After shooting down sixty tanks, he was decorated with the oak leaves of the Knight's Cross of the Iron Cross.

Less formal but still well-earned, a souvenir photo after an Obergefreite of an assault gun crew had received a medal.

Engineer and assault artilleryman load a small flamethrower on an assault gun.

During the heavy fighting for the important Black Sea port of Novorossisk, Assault Gun Unit 210 saw service in August 1942. Here an assault gun covers the advance of a flamethrower shock troop of engineers.

Assault Gun Unit 210 saw service in the Caucasus beginning in early August 1942. At first they took part in the advance toward Rostov; in October they fought with the 1st Mountain Division in the Caucasus Mountains.

On orders from the censors, important details of new assault weapons, such as the muzzle brake, were retouched on photographs (compare photo above). Only then were the photos released for publication.

Further photos of Assault Gun Unit 210 in action, with its emblem, a tiger's head, faintly visible in the center of the bow. The unit was formed at Jüterbog on March 10, 1941 and saw its first service with the Army Group South. These photos were taken in the Caucasus in August 1942.

An assault gun leader of Assault Gun Unit 184 before action. His tasks included leading his gun as part of the ordered action and independently firing on the targets assigned by the platoon leader. In addition, he was to look for targets himself and cooperate with the infantry. Also among the gun leader's tasks was the right choice of firing positions; he supported the aiming gunner, listened to radio messages and oversaw the security of the assault gun.

A platoon leader of the assault guns. If his platoon was assigned independently, he was responsible for cooperation with the infantry, He applied his platoon purposefully, carried out orientation, target spotting and aiming; he chose the assault guns' placement and the first firing positions and thus determined the paths of the two (three as of November 1942) guns of his platoon in attacking. In combat, he had to lead his platoon strictly; he saw to the replacement of losses and to supplying.

The Assault Gun Unit with Three Batteries of Ten Guns Each
(from K.St.N.446a) November 1942

Unit Chief

Staff
(one Assault Gun III)

Staff Battery with Leader's Group
Intelligence Troop
Supply Service
Medical Service
Repair Service
Replacement Group
Rescue Troop
Battery Supply Train
Unit Food Service

1st Assault Gun Battery
with Leader's Group (1 Assault Gun III)
with Combat Battery
Gun Crew (3 platoons with 3 guns each)
Ammunition Troop
Vehicle Repair Troop
with Battery Supply Train

2nd Assault Gun Battery
with Leader's Group (1 Assault Gun III)
with Combat Battery
Gun Crew (3 platoons of 3 guns each)
Ammunition Troop
Vehicle Repair Troop
with Battery Supply Train

3rd Assault Gun Battery
with Leader's Group (1 Assault Gun III)
with Combat Battery
Gun Crew (3 platoons of 3 guns each)
Ammunition Troop
Vehicle Repair Troop
with Battery Supply Train

Total Strength
31 Assault Gun III (Sd.Kfz. 142/1) or 142/2)
1 Ambulance (armored) (Sd. Kfz. 251)
3 18-ton Towing Vehicles (Sd.Kfz. 9)

An important prerequisite for success was close cooperation between the infantry and the assault guns. In practice this included growing problems, because the infantry leaders lacked the necessary knowledge of the nature of the assault gun weapon. Thus in 1942 the platoons of Assault Gun Unit 226 were established in the infantry regiments of the Ninth Army Corps, in order to provide the right basis for antitank defense. As a rule, this soon resulted in the loss of self-propelled gun mounts.

The "Grossdeutsctland" Assault Gun Unit had to endure heavy fighting at Rstrev. Early in January 1943 it was withdrawn and transferred to Smolensk for refreshing. The unit's emblem was a white steel helmet.

The crew of an armored self-propelled mount for a 75.5 cm assault Gun, Type F/8, during a pause in combat in the fall of 1942. At this time, strengthening of the front armor was urgently advocated. In May of that year, Hitler had already advocated a thickness of 80 mm. The added weight of 450 kilograms and the lowered mobility of the assault gun were to be accepted in the bargain. Their future use was foreseen as only to support infantry units.

A gun leader with a shear telescope. The equipment of the assault guns with shear scopes made observation of the battlefield, target recognition and aiming easier.

Until the spring of 1943, most of the self-propelled mounts in the assault gun units still carried the short-barreled assault gun. There were still too few vehicles with the long-barreled gun. Only the increasing production figures (November 1942: 100, and May 1943: 260 assault guns) allowed a general rearmament to take place.

In order to avoid the telltale glitter of shear scopes in snow or bright sunshine, coverings were used, which also were to prevent the viewing openings from filling with ice and snow. This picture from the winter of 1942-43 shows clearly under what complicated weather conditions the assault artillerymen had to fight.

In the winter, the combat readiness of the assault guns was limited by snow and coldness, which particularly affected the mobility of the vehicles.

When one wanted to get the assault guns started and ready for planned action, long periods had to be allowed for starting them. Often hours of laborious work were needed to prepare to march. Important advice was included in the "Manual for the winter use of assault guns", published at Jüterbog in January 1943.

In cold, snowy weather, the use of assault guns in combination with sufficient infantry was regarded as practical for the following types of action: attacks on narrowly limited targets, street fighting, securing of vehicle convoys, defending of localities, major reconnaissance and antitank action.

In harsh winter weather, fighting was often done for shelter in houses and villages. Attacks were thus carried out along streets and with support of the infantry. This picture was taken on the eastern front in the winter of 1942-43. The assault guns, here Type E with the short barrel, were camouflaged with a chalky fluid.

In the "Manual for the winter use of assault guns", it is stated: "When you meet with a Russian T-34, which shows a remarkable mobility even in deep snow, the old assault gun is undoubtedly inferior." It was further recommended that one await the enemy tank in a hiding place and then fire on it with explosive or hollow-charge shells in quick succession.

The new assault guns with the Assault Cannon 40 could match any enemy tank in terms of artillery. As for mobility, they were every bit as limited as the old assault guns. For that reason the tracks were widened with special grippers that were supposed to decrease the ground pressure and improve their grip in ice and snow.

Loading ammunition into an assault gun. With the introduction of the 7.5 cm Assault Cannon 40 L/43, the supply carried aboard the assault guns was cut from 50 to 44 rounds. During production of Type F, the ammunition racks were modified so that 54 shells could finally be carried. These were considerably longer (784.3 mm instead of 523.6 mm) and heavier (11.8 kg instead of 6-7 kg). The steel casings bore the number 6339 St.

From Assault Gun to Pursuit Tank

As early as May 1942, production of the final version of the 7.5 cm Assault Cannon 40, now with the L/48 barrel length, had begun. It was installed in assault guns of Type F and, as of September 1942, in vehicles of the F/8 series. It had a barrel length of 3600 mm (= L/48) and weighed 800 kilograms. It could be ready to fire within twenty seconds. Within a traverse field of 20 degrees, its elevation could extend from -6 to +17 degrees. Shot ranges to 8000 meters were possible. With the Panzer Shell 39 (initial velocity 750 m/s) targets at 1000-meter range could be fired on, without the maximum height of the trajectory exceeding 2.5 meters. In comparison, the height of the most commonly encountered Russian tank,

the T-34, was between 2.45 and 2.65 meters. So it was possible to fire directly and with great accuracy. After 1.44 seconds in the air, the shell struck its target. With the explosive shell (Panzergranate 39), at this distance and with a striking angle of 90 degrees, I could penetrate 96 mm of armor plate. For an angle of 30 degrees, values around 32 mm were still possible. The hard-core shell (Panzergranate 40) penetrated 102 mm under the same conditions, a value that sank to 26 mm of armor plate. Notably more meager performance figures were stated in Army Service Manual 470/20 (Shooting Manual for Tanks, Assault Guns and Armored Scout Cars). To be sure, there were not as good chances of accurate shooting when this type of ammunition was fired, because of its low initial velocity (450 m/s). In general, the effect of the armor-piercing ammunition was better against non-rolled armor plate, such as the turret of the T-34, which was made of cast or pressed material.

The critical situation in the southern part of the eastern front made it necessary, according to a memo of December 21, 1942, to supply Army Group B with, among other things, 52 new assault guns with long barrels. The vehicle shown here has snow tracks, the use of which had to be limited in areas with generally open snowy areas.

Assault Gun III, Type F/8 (Sd.Kfz. 142/1) with the 7.5 cm Assault Cannon 40 L/48, in the summer of 1943. Cooperation with the armored regiments was supposed to be limited to exceptional cases. Within the Panzer divisions, cooperation with Panzer grenadiers was taken for granted.

In addition, the assault gun crews could use the Explosive Shell 34. These shells were made of pressed or cast steel. They had an impact igniter with adjustable delay (0.15 seconds), which allowed the shot effect on the target to be set optimally. For example, when firing ricochet shots, the igniter was set for delay. The shot, landing flat on the ground, bounced off and exploded only when the igniter setting was reached. With ricochet firing, great splinter effect could be achieved against living targets not covered from above.

The assault guns' combat equipment also included 7,5 cm KWK 40 fog cartridges. These fog-producing shells spewed out whitish-gray clouds of smoke when they struck, with a diameter of 30 meters, that lost their effect after 20 to 25 seconds. In 1943, fog cartridges were the smallest part of the primary ammunition supplies in the assault gun units. Ten shells were carried for each assault gun, but as a rule not carried aboard the assault gun. They were kept with the battery's

ammunition troops, which were often equipped as of 1942 with tracked "Maultier" (Mule, Sd.Kfz. 3) trucks. Also among the No. 130 primary ammunition supplies were 130 explosive and 130 armor-piercing shells for each gun. Of these, there were normally 27 of each carried aboard the gun. As a rule, the ammunition racks were removed in order to increase the space for ammunition.

For the assault gun crews, the knowledge of the performance of their weapon and its ammunition was of great importance if they wanted to use it as effectively as possible in action. For example, there were many reminders in their manuals that firing on targets was only practical when a sufficient effect could be expected. Firing on targets outside the effective range of the weapon should be done only when the tactical situation absolutely required it. Thus it has repeatedly been described as incorrect to return Russian tank fire on one's own positions which often came from armored artillery at ranges over 1500 meters. Their front

Development of Armor-piercing Performance for the Primary Armament of German Assault Guns between 1939 and 1945

- All figures are accurate for a range of 500 meters, a striking angle of 60 degrees, and a plate rigidity of 80-100 kg/mm
- Figures for HL (hollow charge) shells apply at all ranges

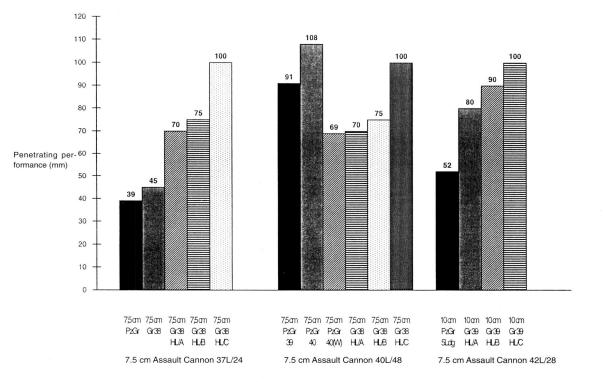

Penetrating performance (mm)

7.5 cm Assault Cannon 37L/24

7.5cm PzGr	7.5cm Gr38	7.5cm Gr38 HL/A	7.5cm Gr38 HL/B	7.5cm Gr38 HL/C
39	45	70	75	100

7.5 cm Assault Cannon 40L/48

7.5cm PzGr 39	7.5cm PzGr 40	7.5cm PzGr 40(W)	7.5cm Gr38 HL/A	7.5cm Gr38 HL/B	7.5cm Gr38 HL/C
91	108	69	70	75	100

7.5 cm Assault Cannon 42L/28

10cm PzGr 5Ldg	10cm Gr39 HL/A	10cm Gr39 HL/B	10cm Gr39 HL/C
52	80	90	100

Production of the armored self-propelled mount for the 7.5 cm assault gun, Type G, began at Alkett in December 1942 and at MIAG in February 1943. By the summer of 1943 these versions were dominant in the assault gun units.

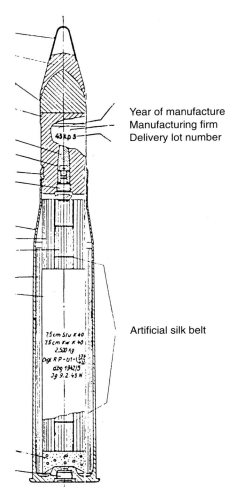

Year of manufacture
Manufacturing firm
Delivery lot number

43 kp 3

Artificial silk belt

7.5 cm Stu K 40
7.5 cm Kw K 40
2.520 kg
Dgl R P - 01 - (378/11)
dbg 1942/5
Jg 9.2.43 N

The 7.5 cm antitank shell—most important type of ammunition for antitank action by the Assault Cannon 40 L/43 and L/48.

In December 1942, production of the Armored Self-propelled Mount, 7.5 cm Cannon, Type G began at the firm of Alkett. The number of sub-contractors increased, and in February the firm of MIAG was included in final assembly. As the manufacturers of the 7.5 cm Assault Cannon 40 L/48, the WIMAG firm of Berlin and the Skoda firm of Pilsen entered the picture. The production figures were constantly increased in expectation of all-out assault gun production. While in January 1943 there were 144 cannons (140 scheduled) made, their production increased to 291 by August of that year. The Army Weapons Office had received a total of 3011 Assault Gun III units in 1943. In 1944 there were 3840. The importance that assault gun production had meanwhile attained in the German armament industry is shown clearly by a comparison with the production figures for the Panzer IV tank, one of the most important German tank models: in the same period there were 3023 and 3125 units made.

The Assault Gun III, Type G used the chassis of the previous Type F/8, but had a wider body with a commander's cupola. The commander's panoramic view was thus improved considerably. This version of the assault gun remained in production until the war ended, though numerous modifications were made in its production in the course of time, partly to improve its capability and armor protection, partly to carry out war-caused practicality measures in terms of saving work forces and materials. As of October 1943, the rotating commander's cupola was eliminated and replaced by an immobile type. Panzer III tank chassis, some of them repaired units from damaged tanks, were included in assault gun production. The production of the Panzer III tank had been concluded in August 1943, ending with Type N (Sd.Kfz. 141/2).

Along with the basic modification of the tank body, the armor thickness in the frontal areas were increased. In the Assault Gun III Types F and F/8, first produced in 1942, the front armor plate had already been [p. 84] thickened to 80 mm.

armor, 60 to 75 mm thick, could not be penetrated at those distances with the armor-piercing ammunition of the assault guns. This amounted to a waste of valuable ammunition. Here too, without troop input, the first faults became apparent. In September 1943, Army High Command 8 reported that the situation concerning ammunition supply for tank guns (also used for assault guns) was extremely tense. For example, to make up for the shortage of explosive shells, the troops were instructed to use hollow-charge shells.

In the first half of 1943, the Army Weapons Office received 1230 assault guns (Sd.Kfz. 142/1) and 119 assault howitzers (Sd.Kfz. 142/2). This almost equaled the year's production of 1942, which had been 1516 vehicles (814 of them with the short-barreled gun). More and more of the older types of assault guns could now be turned over to the replacement army for training purposes. It was still necessary to meet the personnel needs for the hastily built assault guns. This picture of an Assault Gun III Type E was taken at Jüterbog in the winter of 1942-43.

The number of assault guns in the war zone inside Germany was considerable. On March 1, 1945 there were still 446 assault guns and pursuit tanks in the replacement army, as well as fourteen assault guns set up as driving-school tanks.

The Troop Training Camp at Jüterbog, the cradle of the German assault artillery, seen in an air photo taken by the USAAF on May 2, 1944.

The armored self-propelled mount for the 7.5 cm assault gun, Type B, with chassis number 90232, was delivered in the spring of 1941. In 1943 this assault gun was being used by Assault Gun Replacement and Training Unit 400 in Demba, Poland.

Battery training by Assault Gun Replacement and Training Unit 300 in Neisse, Upper Silesia, already established in 1941.

This picture was taken during War Officer Training (KOB-Lehrgang) by Assault Gun Replacement and Training Unit 400 in Demba in 1943. The assault gun, chassis number 90733, is painted sand tan and is an early Type E which had been completed by Alkett in the autumn of 1941.

The same vehicle on a test run. Cooperation between the infantry and the assault gun crews in an attack is being practiced here.

Assault Gun III, Type B, used by Assault gun Replacement and Training Unit 400 in Demba in the early summer of 1943. Here front-experienced officers and NCOs of the artillery were prepared in special training courses for service with assault guns.

NCO Alfred Regeniter, seen during assault gun training at Demba. Until 1943 Regeniter had served as an artillery observer with the heavy unit of Artillery Regiment 255, and had volunteered for assault artillery service. In 1944 he joined Assault Gun Brigade 276 as a lieutenant and became a platoon leader. After shooting down numerous Russian tanks, he was badly wounded early in February 1945. For his service he received the Knight's Cross of the Iron Cross.

Training with the Armored Self-propelled Mount for the 7.5 cm Assault Gun, Type F, at Demba.

The muzzle brake of this Assault Gun III, Type F, is easy to see. Firing without a muzzle brake was forbidden, so as to avoid overworking the recoil brake. Its firm seat had to be checked regularly.

In the autumn of 1943, Assault Gun Replacement and Training Unit 400 was transferred from Demba to Denmark. The unit's job remained the refreshing of hard-hit assault gun crews, the training of replacement crewmen, and the provision of materials ready for use.

Assault Gun Replacement and Training Unit 400 passed through the German capital city in the course of its transfer from Demba to Hadersleben, Denmark. The assault gun shown here, with chassis number 91148, was manufactured in July 1942.

As of May 1943, the first Assault Gun III hulls with 80 mm bow armor were delivered. Older Panzer III hulls were also used, but were fitted with the usual heavier armor.

With the armor plate on the bow and driver's area increased to 80-mm thickness (at 70, 40 and 87 degrees of inclination) and its body armor 30 mm thick and considerably more inclined, the assault gun could withstand the fire of the 76.2-mm tank gun of the Russian T-34 if it did not come closer than 500 meters. At this distance the gun penetrated 69 mm of armor plate. In principle, this was also true of the 76.2-mm Division Cannon 42 (ZIS-3) of the Red Army, which was, though, considerably harder to recognize and fight against. The side and rear armor of the assault gun remained sensitive to shell fire. Another weak spot was the 45-50-mm gun-cradle armor. As of November 1943 it was partially replaced by cast gun-cradle armor, the so-called "sow's head". Disadvantageous results of the strengthened armor appeared in the running gear, which suffered greater wear in the front area.

The assault gun made in 1943, in comparison to that of 1941, was an armored vehicle some two tons heavier with decreased mobility (the specific ground pressure was now 1.10 kp/sq.cm., previously 0.91 kp/sq.cm.). The high-performance weapon and the strong front armor gave it a stronger defensive character, without completely losing its ability to make spatially limited counterthrusts within the parameters of mobile warfare. Out of the typical offensive weapon of the infantry, a defensive weapon had been created within a short period of time. Its main targets were the enemy's tanks.

The importance of the assault gun grew tremendously. There were various reasons for this:

1. The initiative for action was taken by Germany's enemies on all the fronts during 1943. German counterattacks took on more and more of a spatially and temporally limited character, usually to eliminate breakthroughs in their own defenses. With that, the infantry divisions were often overpowered. Often such attacks, made only by the infantry without the support and presence of the assault guns, came to grief.

2. Despite Germany's powerful efforts to produce armaments, the situation here also changed against Germany in 1943. An armaments race

This Assault Gun III, Type E, was set up as a driving-school vehicle for the Assault Gun Replacement and Training Unit 200 in Schweinfurt. For that purpose, the assault cannon was removed. The picture was taken in January 1944.

The vehicle garage of Assault Gun Replacement and Training Unit 200. This unit had been established in Schweinfurt in June 1941. In the spring of 1944 training attained its high point; at this time 5000 assault artillerymen were in training there.

had to be lost by Germany in time, as was well known in the Reich Ministry for Weapons and Ammunition as well as in the other involved military offices. The Foreign Army East Office (II d) reported in 1943 alone that Russia's yearly production numbered 26,500 tanks and assault guns. According to Russian sources, the number was 24,188, of which 16,500 were heavy and medium tanks and 4000 were self-propelled gun mounts. It is no wonder that the tank became the omnipresent and dominant weapon on the battlefield.

3. The German armored forces had to suffer considerable losses in the campaigns of 1941-42. Their new Inspector General, Generaloberst Guderian, set out in 1943 to do everything to reorganize and reequip the armored units. The Panzer troops no longer recovered after their heavy losses in the tank battle around the Kursker Bogen. They could only intervene at particular focal points in the ensuing combat. Thus the infantry units needed the assault guns all the more, as a weapon that equaled the enemy tanks in mobility, fire power and armor protection.

4. The quality of the German infantry in 1943 was no longer comparable to that at the beginning of the eastern campaign in June 1941. Their lacking numbers and offensive or defensive power could be seen as a German weakness already in July 1943 during the battle of Kursk. The assignment of additional assault guns was to help make up for this loss of quality. At this point we can recall an old military experience from the time of the Silesian Wars in the 18th century: The worse the infantry, the more important the artillery.

An armored self-propelled mount for the 7.5 cm assault gun, Type G. The widened armored body and the commander's cupola can be seen clearly. As of May 1943, the first assault guns with 80 mm bow armor were delivered. [end of file, page 86]

Noticeable features of these Assault Gun III, Type G vehicles are the narrow armored aprons on the sides. They were photographed in the summer of 1943.

In December 1941, it was already being urged that the assault gun be equipped with a weapon whose ammunition could produce a greater explosive and splintering effect than was provided by the 7.5 cm assault cannon. At the end of 1942, series production of the 10.5 cm assault howitzer began. As of February 1943, it was delivered, along with the Type G Assault Gun III. The gun had been developed from the 10,5 cm light field howitzer by the firm of Rheinmetall. Unlike the assault cannon, the Assault Howitzer 42 had to be loaded with separated ammunition, first the shell, then the cartridge case. The technically possible maximum range of 10650 as compared with 12325 meters played a smaller role in practice (thus the muzzle brake could be done away with as of September 1944). It was much more important to take battlefield targets up to 2000 meters away under effective fire with the 15.55-kilogram howitzer shell (1.38 kf explosive). Effective splinters flew up to 40 meters to the sides and 10 meters forward when the shell exploded. The advantage becomes clear when one compares these figures with those of the 7.5 cm Explosive Shell 34 of the 7,5 cm Assault Gun 40. The shell weighed

5.74 kilograms, and contained 0.68 kg explosive. The splinter effect extended up to 15 meters to the sides and seven meters to the front.

To attack armored vehicles, every assault howitzer carried ten hollow-charge shells and cartridges in its combat supplies, along with 26 explosive shells. The penetrating power of the 10 cm 39 Red/HL/B shell against armor plate was stated at 90 mm, that of the HL/C at 100 mm: a sufficient achievement that could be attained irrespective of the range. The crews of the assault howitzers nevertheless saw themselves confronted by a whole series of problems in combat with enemy tanks. Because of the low muzzle velocity of the hollow-charge shell (495 m/s), the range for direct firing extended from 500 meters down. There was also the slow rate of fire, of three to five shots per minute. Fire and smoke production in firing were greater than those of the 7.5 cm assault cannon, which meant that the assault howitzer's position could be spotted much more quickly. At distances under 500 meters, as was already noted elsewhere, the front armor of the assault gun offered no protection from the fire of tank guns.

The Assault Gun III, Type G, in comparison with the assault howitzer. This vehicle, with chassis number 95219, was manufactured by the MIAG firm in May 1943. It was presumably lost in Italy, for it was one of the first assault gun of this type that was evaluated by the Allies (see text on armor plate).

Side view of Assault Howitzer III, Type G. Essentially, it can be told apart from the assault gun armed with the 7.5 cc Assault Cannon 40 only in the shorter, more compact barrel of the 10.5 cm Assault Howitzer 42. Assault howitzers were manufactured only by Alkett; their share in assault gun production was retrogressive. In all, 1299 of them had been mounted by 1945.

Effective March 2, 1943, the inclusion of assault howitzers in assault gun batteries took place according to War Strength Document 446 a of November 1942. They could now be equipped with seven 7.5 cm Assault Gun 40 (Sd.Kfz. 142/1) and three 10.5 cm Assault Howitzer 42 (Sd.Kfz. 142/2), or as was customary before, only with assault guns. At the beginning of the battle in the Kursker Bogen, assault howitzers were on hand in nine assault gun units, and eight of them were used directly in the combat. The equipping and distribution of assault howitzers sometimes varied very much, as shown in a report about Assault Gun Unit 261 on September 6, 1943. It had been newly established in Schweinfurt and Altengrabow and transported to Poltava. With a total strength of 31 vehicles, there were in its

1st Battery seven 7.5 cm Assault Gun 40 and three 10.5 cm Assault Howitzer 42, in the
2nd Battery ten 7.5 cm Assault Gun 40, and in the 3rd Battery nine 7.5 cm Assault Gun 40 and one 10.5 cm Assault Howitzer 42.

This was a fighting force with which the unit commander, Hauptmann Kokott, could not be very happy. Assault gun units brought more than twenty assault guns into combat together only in rare instances.

The production of the Assault Howitzer 42 experienced strong variations. In May 1943, 45 assault howitzers were delivered, in August of that year only five (there should have been thirty). In May 1944 there were 46. As early as mid-October 1943, the Reich Minister for Weapons and Ammunition, Albert Speer, urged a new distribution of assault howitzer and assault gun production, based on available reports from the fronts. The assault howitzer, since it was only conditionally suited for combat against tanks, enjoyed less and less popularity. Thus it is not surprising that in December 1944, 452 Assault Gun 40 units were delivered, compared with only forty of the Assault Howitzer 42.

The numbers of assault gun forces were increased in 1943. In July of that year there were 26 assault gun units and two batteries on the eastern front. In Army Groups Center and South, Assault Gun Units No. 177, 185, 189, 228, 244, 245, 904, 905, 909 and 911 and the "Grossdeutschland" assault gun unit were drawn into the vigorous fighting around the Kursker Bogen. In all, there were 533 assault guns in action here, representing 55 percent of the eastern forces' total. At this time, assault guns had already spread beyond the assault gun forces. They were now to be found in numerous other units. The divisions and brigades of the Waffen-SS generally had their own battery, sometimes even an entire unit. A similar development was seen in several divisions of the Luftwaffe, which were also given their own assault gun units. In the Army there were other divisions with their own assault gun unit or battery. In addition, the "Grossdeutschland" Panzer grenadiers and the 18th Artillery Division (Assault Gun Battery 741), established in October 1943, must be noted.

Falsch   Richtig

For driver training, right and wrong illustrations were often used by the replacement army. The example here shows how a tree stump should be driven over by an assault gun.

The fitting of armor-plate aprons on the sides of newly built assault guns began in April 1943. In May, 330 reequipping sets were delivered to the eastern front. The troops complained about the insufficient attachments. Often the aprons were lost in combat.

In the morning hours of July 5, 1943, assault guns rolled through the first holes broken in the Russian defense lines by attacking troops, in order to be ready for further combat. In all, ten independent assault gun units were available, while others were on hand with several Panzer and Panzer grenadier divisions of the Army and Waffen-SS.

In the course of Operation "Citadel" in July 1943, the German attack came to a stop in the deep defenses of the Red Army. In part, 1500 tanks and 1700 anti-personnel mines were transferred per front kilometer; 92 artillery guns and grenade launchers were in one kilometer in the main sectors. Guns of the corps artillery, like these destroyed 152 mm cannon howitzers, were also put to use to fight off German tanks with direct fire. In the background is an Assault Gun III, Type G.

In the 4th Panzer Army and the Arly Section Kempf in the Kursker Boden, where 202 assault guns had been reported as ready for action on July 5, 1943, there were 14 total losses by July 16. The readiness for action in the assault gun units amounted to only 40 percent, though, after only a few days of action. This situation was mainly attributable to technical breakdowns. Only with the beginning withdrawals did the losses increase.

An Armored Self-propelled Mount for Assault Gun 7.5 cm Cannon, Type F. These vehicles were still quite numerous in the assault gun units in 1943. There were also a few assault guns of older series that were rearmed, in the process of general maintenance, with the 7.5 cm Assault Cannon 40.

Action photos from the summer of 1943. In a training manual it is said about the use of assault guns: "To the last man, the slogan of the infantryman must also apply to the assault guns: 'Forward and at the enemy!' Assault guns are an offensive weapon and give the infantry immediate, powerful support through mobility, fire and striking power."

On this assault gun, the bow armor of the hull or the front armor of the upper body were increased to 80 mm by screwed-on plates 30 mm thick. Vehicles of this type were delivered through October 1943 (hull) and June 1944 (body).

In another manual for the assault artillerymen it was said of the gun's equipment: "Fasten all equipment according to tested plans. Never change positions without reason. . . ." In practice it was often different, as this picture of an overloaded assault gun shows. Often the ammunition attachments were removed in order to be able to carry a larger supply than the usual 54 shells. Many objects found places on and in assault guns if they served to lighten the heavy burdens of the soldier: provisions, clothing, cooking utensils, blankets and tent canvas.

The assault gun crews always had to reckon on the possibility of encountering enemy tanks. It was often a case of being the first to open fire at as close a range as possible in order to have an immediate effect. The swift course of the shells that could be achieved with the long-barreled guns demanded a free field of fire. At ranges up to 900 meters, firing over one's own infantry was not allowed because of the low trajectory.

Great importance was given to the camouflaging of assault guns and their surprise attacks: "The more surprising the approach of the assault guns is, the greater their effect is. If the enemy knows that assault guns are on hand, he will expect an attack and strengthen his defenses."

Victor and vanquished on the battlefield. In the course of the war, the assault guns caused serious losses to the Russian armored troops and held these enemies at a respectful distance in heavy combat. A negative influence on the course of the war was the growth of Russian tank production to truly incredible numbers (24,188 tanks in 1943 and 28984 in 1944).

The many antitank guns and field cannons of the Red Army, hard to recognize when moved into position well camouflaged, were a greater danger to the assault guns. For example, one antitank brigade had seventy-two 57- and 76-mm cannons. Here, to be sure, the assault gun was the victor in a duel with a Russian 76 cm division cannon, 1942 model (ZIS-3).

The Cure-all Assault Gun

Beginning early in 1943, Panzerjäger units of various divisions were given their own assault gun units. The Tenth Army Corps received a message on September 10, 1943, stating that personnel units of one Panzerjäger company (mot.Z.) each of the 6th and 7th Infantry Division had to be transferred to Panzerjäger Replacement and Training Units 1 (Allenstein, East Prussia) and 7 (Munich) for restructuring, refreshing and rearming. The 2nd Company of Panzerjäger Unit 7, now an assault gun unit, returned to its division at the end of 1943, after having obtained its vehicles from the Army Arsenal Office in Magdeburg. Confusing in several ways was the term "Assault Gun Unit". First of all, its strength corresponded to that of an assault gun battery. Second, the Panzerjäger units in the infantry, rifle, mountain rifle, Panzer grenadier and later Volksgrenadier divisions were not subordinate to the artillery, but to the armored troops. As a result they were given the designation of "Jagdpanzer Company". For that reason, and because their structure and their primary functions differed from those of the assault gun units, they are not treated and described more fully here. From a technical standpoint, the assault guns amounted only to a temporary solution; in 1944-45 they were replaced in the Panzerjäger units of the infantry divisions by the Jagdpanzer 38 "Hetzer". In the Panzer and Panzergrenadier divisions, the Jagdpanzer IV was put into service.

The use of assault guns in the Panzer units of some Panzer and Panzer grenadier divisions of the Army and the Waffen-SS is another interesting theme that, for the reasons already cited, cannot be portrayed more fully. The main reason must be recognized as the lack of tanks that became more obvious at the beginning of 1943. Generaloberst Heinz Guderian attempted at this time to reorganize the Panzer troops. He planned to absorb all the artillery's assault gun units into the Panzer troops. This, as Guderian writes in his memoirs, had a good reason, for ". . . the production of assault guns formed a considerable part of tank production". With the gradual blending of assault gun and Panzerjäger units, the potential performance of the hard-pressed antitank forces in particular was supposed to be increased. What the Inspector General of the Panzer Troops did not write was the fact that such a strong concentration on the production and use of all armored vehicles under the aegis of the Panzer forces would have led to a further rapid degeneration of the fighting value of the infantry forces themselves. This could not be accepted as such in the interests of the further conduct of the war. The Führer's decision of March 13, 19433 ended this disagreement. "The assault guns remain as before with the artillery and have the purpose of serving the infantry further as armored escort artillery. As long as assault gun units are turned over to Panzer divisions, they remain for the time being under the direction of the Inspector General of the Panzer Troops." Guderian could not put his plans into operation. What he achieved was that, out of the steady monthly production as of May 1943, 100 assault guns were turned over to the Panzer troops, in order to equip the armored units of Panzergrenadier and Panzer divisions as well as special units. That was also supposed to be a temporary solution, as with the Panzerjägers. Panzer companies that were equipped with assault guns were formed into three or four platoons with totals of 14 or 22 vehicles. The III./Panzer Regiment 24 (24th Panzer Division), for example, received two squadrons with Assault Gun III vehicles. In the 14th Panzer Division, the complete second unit of Panzer Regiment 36 was equipped with them, and in the 16th Panzer Division the third unit was.

Despite numerous advantages of the assault guns, which could be used advantageously, especially in defense, when used correctly, the equipping of some Panzer units with assault guns did not blend with the nature of the tank weapon. This can be seen in numerous action reports. Yet it was practiced until the war ended.

This type of utilization did not remain without influence on the further technical development of the assault gun. This could be seen in the assault gun units with the clearly growing tendency to push the assault guns more and more strongly into the role of an infantry escort tank. To suit this manner of action, it was necessary to equip the assault gun along the way with weapons that would make it possible to fight effectively against the enemy infantry (and not just close combat against tanks). The machine-gun shield in front of the gunner's hatch, introduced already in 1942, no longer met these requirements, for it offered only insufficient protection. Therefore in 1944 an all-around machine gun, to be operated under armor protection, went into troop testing. This particular close-combat defensive weapon was introduced in May 1944, but was seldom to be seen among the assault guns. And finally, mounting a coaxial machine gun in the gun cradle armor was begun in

1944. The troops had already requested this change in 1943. By now, all the assault gun lacked was a rotating turret.

Important planning principles for the production and use of assault guns can be seen in the position taken by the Inspector General of the Panzer Troops (Abt.Org.Nr. 2190/43 g.Kdos) on December 18, 1943 and presented to the armament study of the Army General Staff for 1944. Among other things, the urged simplification of types was accepted, but according to it, the Assault Gun IV, not the Assault Gun III, was to be included among the major types. Its production was begun in December 1943 by the Krupp-Gruson firm in Magdeburg, in place of the Panzer IV tank. In his statement, the Inspector General called for an average monthly production of 650 assault guns. They were to be divided as follows:

1.   For the equipping of ten Panzerjäger companies in the infantry divisions per month—contingent of 100 assault guns.

In September 1943 a company of Panzerjäger Unit 7 (7th Infantry Division) was equipped with ten assault guns. The vehicles were obtained from the Army Arsenal Office in Magdeburg. Their later fitting with armored track aprons and additional track links on the bow of the hull was done by the troops. The vehicle shown here bears number 212 (for the second company of Panzerjäger Unit 7, first platoon, second vehicle).

Assault guns in the Panzer troops—here with Panzer Unit 5 of the 25th Panzergrenadier Division. The panzer unit joined this division in the autumn of 1943 and was structured with three companies of 17 assault guns and a staff with three Armored Command Car III vehicles. It was wiped out in the pocket of Minsk.

Assault guns were to be used only in exceptional cases to equip individual Panzer units of Panzer and Panzergrenadier divisions. Thanks to their low bodies, they were less vulnerable to enemy fire. On the other hand, they were more vulnerable in close combat and less independent in combat with enemy infantry. Their use was to take place, if possible, only in conjunction with Panzergrenadier troops

The equipping of assault guns with optical devices among the Panzer troops did not differ from that of the vehicles of the assault artillery. Panoramic Telescope 32, SFL Targeting Scope 1a, and Shear Scope SF 14 Z were used.

The use of assault guns in tank units, according to an experience report of the III./Panzer Regiment 36 (14th Panzer Division) of December 1943, showed the following disadvantages in combat: To fight targets appearing to the left or right, the assault gun always had to turn its front end toward the enemy; this slowed the rate of fire and decreased mobility. And it lacked a machine gun to be utilized under armor protection.

An Assault Gun III, Type G, of Panzer Regiment 33 (9th Panzer Division). This regiment had already appeared in the east in 1943 with its own Panzer companies of 14 assault guns. In order to make up for shortages in tank production, one company of the I./Panzer Regiment 33 was newly equipped with them in December 1944 and saw service in the west.

2. For the establishment and completion of 40 assault gun units of the artillery, with 45 assault guns each—monthly contingent 200 assault guns.
3. For the completion of all Panzer units of the Panzergrenadier divisions.
4. For the completion of a Panzer unit (FKL) and seven Panzer companies (FKL).

Just the commanded establishment of 120 Jagdpanzer companies (assault gun or Panzerjäger companies) of the infantry added up to an actual need for 2040 assault guns and light tank destroyers. According to a priority program of the firms that built armored vehicle bodies in 1944-45, the Assault Gun III was to be delivered until November 1944 at the rate of 500 units and the Assault Gun IV unit, March 1945 at the rate of 100 to 130 units. The series production of both models could then run out in May 1945 with 50 units, or in June of that year with 100 units. That was what the planners envisioned. Reality looked different. While there were still 365 Assault Gun III and 80 Assault Gun IV units received by the Army Weapons Office in November 1944, this number decreased by April 1945 to 95 Assault Gun III and three Assault Gun IV units (statistic include the numbers of assault howitzers).

In April, May and June of 1943, a total of 61 assault guns were assigned to radio-controlled tank companies. According to the War Strength Directive 1171 f, of January 1, 1943 and June 1, 1944, each company had ten assault guns as guiding tanks and 36 explosives carriers (Sd.Kfz. 301).

The assault gun company of the Rhodos Panzer Unit, similarly to the assault gun battery of Special Unit 287, received assault guns equipped for the tropics. They were organized as a battery of seven guns and intended for service with the German-Italian Panzer Army in North Africa. The picture shows an assault howitzer.

The complicated road and path conditions in the east led in many cases to the loss of heavy weapons and vehicles. Their rescue was often difficult or failed for lack of suitable towing equipment. This series of pictures shows three phases in the removal of an assault gun that had broken through the ice that covered a river.

Towing an assault gun that ran into a Russian minefield and suffered damage to its tracks when an antitank mine exploded. Towing by an assault gun was not encouraged because of the overburdening of the powerplant, but was a common practice, especially in separated action of batteries. The towing staffs and repair-shop services could not always guarantee orderly servicing of the assault gun batteries in such cases.

In Assault Gun Unit 190, which was redesignated an assault gun brigade in October 1943, the repair services had their hands full to make damaged self-propelled guns ready for service again. The 1st Battery alone lost half of its vehicles on November 8 when it struck a German minefield during its withdrawal.

Fitting Into a Changed Organization

On February 1, 1944 a new War Strength Directive (K.St.N. 446b) went into effect. According to it, an assault gun battery consisted of four platoons, one of which had three 10.5 cm Assault Howitzer 42 units. Three platoons were each equipped with three 7.5 cm Assault Cannon 40 units. Together with the two assault guns of the battery leader, that made fourteen vehicles in each battery. The unit with three batteries had 45 assault guns, including the three of the staff. Older war strength schedules retained their validity.

On February 25, 1944 the reorganizing of the assault gun units into assault gun brigades was ordered. Almost parallel to it, the establishment of assault gun escort batteries was begun in February. They were to consist of particularly battle-experienced infantrymen and be formed for cooperation with the assault guns. Their tasks consisted of accompanying the assault guns in combat and their protection or defense. They were to guarantee the better utilization of the offensive power of assault guns and increase the enthusiasm of action and attack. Measures such as the forming of escort batteries are indicative of the infantry's declining performance, which had meanwhile become obvious. The assault gun forces, directed to cooperate with the infantry, had to accept losses more and more often that were attributable to poor cooperation and lack of understanding of their needs.

The previous assault gun batteries were charged with the organization of escort batteries. In addition, in October 1944 a similar reorganization of the Jagdpanzer (assault gun) companies of the Panzerjäger units of the infantry divisions was undertaken. They were given grenadier escort platoons. While such structuring got results, no long existence was guaranteed the escort tank batteries (according to K.St.N. 447 Addenda) with twelve Panzer II (2 cm) (Sd.Kfz. 121). They were disbanded in August 1944.

In June 1944, the equipping of the brigade staffs with assault guns was regulated uniformly. With that, the total strengths for assault gun brigades was 31 or 45 assault guns. In June 1944 there were 48 assault gun and three assault artillery brigades available, 33 of which saw service in the east. At this time, twelve brigades were in Germany being refreshed. This apportioning resulted in clear focal points of action for the assault artillery that was in the east. There essential combat experience was gained in hard defensive fighting, which helped to determine their organization, training and basis for action, and at the same time resulted in constant changes.

Wedge Shape

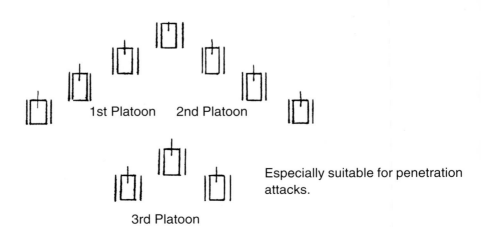

1st Platoon   2nd Platoon

3rd Platoon

Especially suitable for penetration attacks.

The Army Assault Artillery Brigade
with three batteries of 14 guns each
(according to K.St.N. 446B) June 1944

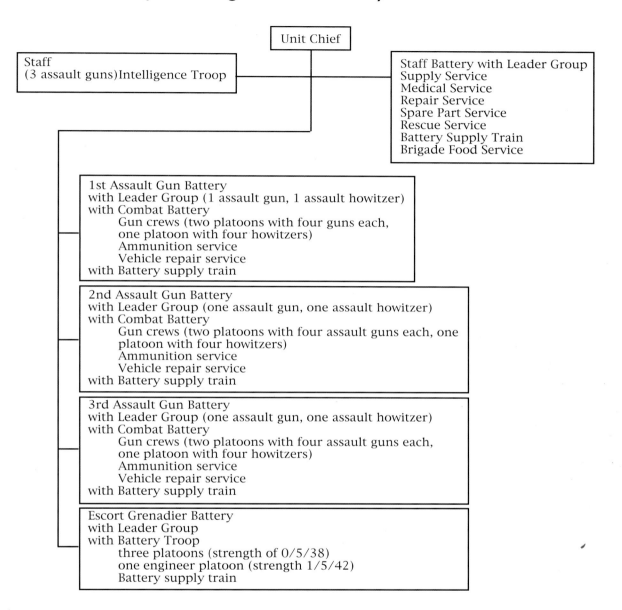

Unit Chief

Staff
(3 assault guns)Intelligence Troop

Staff Battery with Leader Group
Supply Service
Medical Service
Repair Service
Spare Part Service
Rescue Service
Battery Supply Train
Brigade Food Service

1st Assault Gun Battery
with Leader Group (1 assault gun, 1 assault howitzer)
with Combat Battery
    Gun crews (two platoons with four guns each,
    one platoon with four howitzers)
    Ammunition service
    Vehicle repair service
with Battery supply train

2nd Assault Gun Battery
with Leader Group (one assault gun, one assault howitzer)
with Combat Battery
    Gun crews (two platoons with four assault guns each, one
    platoon with four howitzers)
    Ammunition service
    Vehicle repair service
with Battery supply train

3rd Assault Gun Battery
with Leader Group (one assault gun, one assault howitzer)
with Combat Battery
    Gun crews (two platoons with four assault guns each,
    one platoon with four howitzers)
    Ammunition service
    Vehicle repair service
with Battery supply train

Escort Grenadier Battery
with Leader Group
with Battery Troop
    three platoons (strength of 0/5/38)
    one engineer platoon (strength 1/5/42)
    Battery supply train

Total Strength
0 Assault Gun III (Sd.Kfz. 142/1)
or IV (Sd.Kfz. 163)
15 Assault Howitzer III (Sd.Kfz. 142/2)
1 Ambulance (armored) (Sd.Kfz. 251)
3 Towing tractors, 18-ton (Sd.Kfz. 9)

Note: Assault gun units were also structured, as of June 1, 1944, with three batteries of three platoons with three assault guns, whereby the second platoon always were supposed to have three 10.5 cm Assault Howitzer III (Sd.Kfz. 142/2) (K.St.N. 446 A).

A weak point in the added armor plate of the Assault Gun III, Type G, repainted the welded, box-shaped gun-cradle armor with a thickness of 45 to 50 mm. It remained in production, for reasons of manufacturing technology, even after the introduction of the cast (sow's-head) gun cradle armor.

Guns with cast gun cradle armor were produced as of November 1943. The so-called "sow's-head shield" offered better ballistic protection, enabled large-scale manufacturing with a considerable reduction of work expenditure, but had to be made 20 to 30 percent thicker then the shield of homogeneous welded armor plate in order to have comparable strength when fired on.

The Assault Gun in Battle

The evaluation which had been attributed to the assault gun as a tank destroyer remained unchangingly high. In Memo 9/12 g "Temporary Guidelines for the use of antitank weapons in defense" of May 20, 1944, it was said: ". . . Assault guns have proved to be the best weapon for fighting against tanks to date." In a directive of the Inspector General of the Panzer Troops (Abt.Ausb. IV(PzJäg) No. 8950/44 geh.) of June 12, 1944, the role of the assault guns was characterized as follows: ". . . closed action for pursuing enemy tanks, preceding an infantry attack, armor-piercing shock reserve in defense". Statements of principle about the use of assault guns could also be taken from Army Service Directive 298/3a "Command and Combat of the Panzer Grenadier, Book I" The Panzergrenadier Battalion (armored)" of August 5, 1944. "Assault guns are especially suited to bring the Panzergrenadiers to the enemy through their shock- and firepower and to achieve the breaking of the enemy's main battle line as well as advancing through the main battlefield." In this it needed constant protection from the Panzergrenadiers. This very cooperation between the assault guns and the infantry often, as already mentioned in a different context, left something to be desired. Help was to be provided by Directive 18/10 "What must the grenadier know about assault guns and tank destroyers?" Much scope for the use of assault guns against enemy tanks was also provided in this document. "When enemy tanks arrive, then assault guns and tank destroyers plunge toward them, no matter what task they otherwise have been assigned," it is stated on page 7.

Great importance was placed on shooting in training for the assault artillery. For the fastest possible destruction of the target, along with the right application of the unit, the battery or the individual assault gun, the choice of the shooting procedure that would destroy the target most quickly and the mastery of the firing rules were of decisive importance. Vital to the choice of the firing procedure to be used and the type of ammunition were: 1. the combat mission, 2. the type and position of the target, 3. the distance, terrain and trajectory, and 4. the prospects of striking the target.

The choice of a direction of attack that would benefit the firing (flanking effect, consideration of the sun's position, wind direction and dust development) should always be of importance. As a rule, it was a task of the battery leader to direct the fire by radio. Firing was done out of the firing position. Against tanks, the antitank shell was to be used at ranges under 1200 meters and the hollow-charge shell under 600 meters, to begin immediately with effective shooting, as stated in Army Service Manual 470/20 "Panzer Firing Instructions (Firing instructions for tanks, assault guns and armored scout cars)" of January 8, 1944. At greater distances, effective shooting always had to be preceded by shooting-in. When the range was then determined, then a unification of fire was to be commended at once, so as to achieve the effect more quickly.

The troop emblem of Assault Gun Brigade 177.

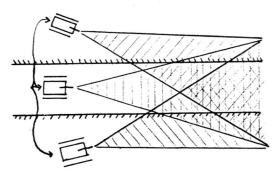

3. Starker frontaler Wider-
stand, volle Entfaltung der
auf dem Wege behinderten
Gefechtskraft. Zangenförm.
Ansatz d. Sturmgeschütze.

The wealth of technical and tactical experience gained from the front action of the assault gun units resulted in a multitude of manuals and memos that appeared at short intervals in 1944-45. Sometimes simple sketches served to illustrate advantageous tactical formations for conducting firefights.

Increasing importance in the use of the assault gun was given to firing at night. To be sure, poorer vision and unfavorable chances of observation limited the possible effectiveness of the weapon, but that was also true for the enemy. With the right use of forces and a successful surprise effect, night actions offered good possibilities of success. Thus it is not surprising that Memo 47a/ 22 "Firing at Night" was issued on January 1, 1945. Along with natural light (full moon), various means, such as the Parachute Light Cartridge 41 and the Gun Parachute Light Cartridge, were to be used to light up the forefield. If necessary, grenadiers operating ahead of the assault guns could light up targets lying farther forward for the assault guns by "shooting and setting wooden houses, heaps of straw or hay afire, and by firing incendiary ammunition." In addition, a special firing searchlight for assault guns and howitzers had been introduced, with a range of 600 meters. Despite complicated conditions for action, the use of assault guns at night by their units, which like the Panzer units suffered under oppressive Allied air superiority on all fronts, afforded greater possibilities for action and success.

Extensive reports and, sometimes, detailed statistics on the success of the assault guns in action are available. Although they are not all to be found in the realm of antitank defense, the numbers of shot-down tanks is cited again and again as an important statistic by which to evaluate the success of this weapon. It can be seen from them that, by the time World War II ended, they had wiped out well over 20,000 tanks. Russian tank soldiers had instructions to break off the combat on the arrival of German assault guns, get around them on the flanks if possible, and outmaneuver them.

Individual experiments with antitank fire from assault guns led to remarkable results. In Italy in May 1944, the 10th Army had a total of 118 assault guns, including, to be sure, 56 Italian captured assault guns that were only marginally usable in combat against tanks. Of the 432 reported tanks destroyed in this army's sector, 33 were scored by the assault guns. In this period 99 assault guns were lost, 50 of them vehicles made in Italy. A good part of the high losses resulted from vigorous Allied air raids.

In February 1945, the 9th Army shot down 49 tanks on the eastern front, almost 50 percent (24) by assault guns and pursuit tanks. Only three tanks were shot down by Panzer IV and V tanks. The high percentage of tanks shot down by assault guns and pursuit tanks was only partly attributable to the fact that the number of these vehicles had meanwhile exceeded that of the tanks considerably. In January 1945 there were around 3500 tanks and some 4300 assault guns on hand. Among them were captured assault guns, plus the still goodly number of 225 assault guns with the short-barreled 7.5 cm cannon. The latter were found chiefly in the replacement army, where there were 446 assault guns and pursuit tanks at the beginning of March 1945.

What advantages assault guns possessed in comparison with tanks in firefights with enemy tanks can be seen clearly in this picture. Their low height allowed better utilization of terrain conditions and vegetation and made camouflage easier. In addition, assault guns were faster and easier to manufacture. Their price was 82,500 Reichsmark per unit, while a Panzer III tank cost 96,183 Reichsmark.

The gun leader carefully observes the field of battle through the SF 14 Z aiming telescope, so as to inform the gunner of targets to fire on as soon as they appear. Terrain examination and reconnaissance could not be interrupted during the course of the battle. The Double Earphones b (replaced in 1944 by Radio Headset a) worn by the commander can be seen clearly.

Among the most important assembly factories for assault guns was the Altmärkische Kettenwerk GmbH (Alkett) in Berlin-Tegel. Here 2520 of the total of 3840 assault guns (Sd.Kfz. 142/1) built in 1944 were made. In addition, 903 assault howitzers (Sd.Kfz. 142/2) were produced.

Alkett manufactured assault guns in their Borsigwalde, Spandau and Falkensee works, which were often the targets of Allied bombing attacks. This picture of the final assembly was taken in the spring of 1944.

Preparation for acceptance by the Army Weapons Office. The second vehicle in the lower part of the picture is an Assault Howitzer III, Type G (Sd.Kfz. 142/2), which was produced only by Alkett. The picture dates from March or April 1944.

Assault gun production at the Borsigwalde Alkett factory in the summer of 1943. All the assault guns were built with the boxlike gun cradle armor; the Assault Howitzer III (fourth vehicle from right) has a muzzle brake.

The firm of F. Krupp Grusonwerk AG in Magdeburg-Buckau, shown in an Allied air photo taken on July 8, 1944. The production of the Assault Gun IV (Sd.Kfz. 167) was begun here in January 1944.

For the Assault Gun IV, the armored upper body of the Assault Gun III could be used almost unchanged. The Gruson works were bombed five times in all in 1944-45, without seriously disturbing the final assembly. In 1944 1006 Assault Gun IV units were delivered from there, in 1945 another 105 (the last three in April).

The completion of the Assault Gun IV became considerably more difficult in the last months of the war, influenced by the destruction of railroad connections. The upper bodies came from the Brandenburgische Eisenwerke in Brandenburg on the Havel. Boehler of Kapfenberg produced the hulls, as did the Oderndau Iron Works in Linz and Krupp in Essen.

At the Limits of Performance

At the end of 1944 it became more and more obvious that the assault gun, great numbers of which were then with the troops, could no longer satisfy all its requirements as a vehicle or a weapon. The 7.5 cm Assault Cannon 40 L/48 had been introduced more than two years before. The armored troops of Germany's enemies had undergone considerable development during this period. This applied to technology, numbers of tanks, and methods of using them. In addition, effective antitank guns had been introduced in all armies; the infantry likewise obtained antitank weapons with great penetrating power. And the constant threat from fighter-bombers from the air should not be forgotten.

On the western front, the Americans used their most important tank type, the M4, and its derivatives (the Sherman), more and more of them armed with high-performance 76 mm tank guns. This weapon, using antitank shells, could penetrate 88 mm of armor plate at a range of 914 meters (with a striking angle of 60 degrees). The HVAP shell even penetrated 133 mm. In the frontal area, armor plate 60 to 100 mm thick protected the tank. A few models carried armor 135 to 150 mm thick.

The British armored troops put various tank models with the 17-pounder (76.2 mm) tank gun into action; it could penetrate 120 to 186 mm of armor plate at 475 meters.

An excerpt from a report, dated February 16, 1944, about Assault Gun Brigade 239, was typical of the use of the assault gun in the last third of World War II: "A breakthrough of several T-34 tanks into the extremely important town of Shilki was cleaned up by NCO Dienemann with his gun, shooting down five tanks, after the two battalions applied there evacuated their positions in flight."

The American M4A1 (76 mm) HVSS tank with its 76 mm M1A1 tank gun. This vehicle, completely covered with cast steel armor (65 to 85 mm thick), appeared at the front in greater numbers as of the end of 1944. The tank weighed 42 tons and reached 38 kph.

The Russian T-34/85 tank, armed with an 85 mm tank gun, also weighed 32 tons. Along with its high speed (53 kph), it caused many a surprise with its outstanding off-road capability (specific ground pressure 0.83 kp/sq.cm.— compared with 0.93 kp/sq.cm. for the Assault Gun III, Type G).

Less mobile and also more ponderous in its rate of fire, the IS-2 heavy tank had a 122 mm tank gun. It was utilized in the heavy armored penetration regiments of the Red Army. This picture was taken in Altdamm in the spring of 1945.

In this picture, which shows an Assault Gun IV (Sd.Kfz. 167) in a readiness area, several features can be seen. The gun shield for the machine gun by the loader is missing and the right front of the body has been strengthen with concrete. The driver's balcony on the left side has an angled armor plate for additional protection. The "Zimmerit" protective coating to keep off magnetic hollow charges was omitted beginning in September 1944.

In the east, where most of the assault gun and army artillery brigades saw service in costly defensive fighting, the Russian tanks likewise met them with great numbers of improved and new tank models. Above all, the T-34/85 medium tank, armed with an 85-mm ZIS-S-53 tank gun, must be named. At 500 meters, 89 mm of armor plate could be penetrated, and at 1500 meters 74 mm (at a 60-degree angle). On the Tank Recognition Card, as of November 1, 1944, this type of tank was called the "most important and most often encountered tank of the Soviet Russian armored forces. . . . The larger gun with a longer barrel than that of the previous T-34 allows this tank to wage armored warfare at long range. . . ." Armor plate 60 to 76 mm thick protected the front of its hull and turret. In addition, of the 14,773 T-34 tanks that were delivered in 1944, more than 11,000

were of this new type. Here are comparison figures: In that year, Germany attained a yearly production of only 18,284 tanks, assault guns, pursuit tanks and self-propelled gun mounts together.

The armored troops of the Red Army received a further development in 1944—that of the heavy IS ("Stalin") tank. Its tank gun had a caliber of 122 mm and fired an antitank shell that penetrated 126 mm of armor plate at 1000 meters; even at 2500 meters it could still penetrate 90 mm. At this range a direct hit with the 25-kilogram shell could fully destroy an assault gun. On the other hand, the front armor, 100 to 135 mm thick (90 too 100 mm on the sides), offered reliable protection from the shells fired by the 7.5 cm assault gun; the Antitank Shell 39 could not penetrate it even at 100 meters. The 7.5 cm Antitank Shell 40 (100 m/60 degrees/126 mm) was only rarely available in 1944.

Comparing the statistics makes clear that the assault gun crews had a rough time in combat against tanks. For them it was important in a confrontation with "Stalin" tanks to utilize the advantages of the assault gun, the low body and the greater mobility, in order to make it harder for the Russian tank crews to score direct hits. The latter had to load their guns with two-piece ammunition (shell and cartridge), which kept their rate of fire low. It was some two or three rounds per minute. With combat-tested, experienced crews on the assault guns, they could use clever driving maneuvers and accurate shooting to gain success with their shots.

Leutnant Alfred Regeniter of the 3./Assault Gun Brigade 276 described his first encounter with the heavy Russian tank as follows: "On the afternoon of October 21, 1944 the battery made an attack in the Steinkrug area of East Prussia. But we were scarcely able to get out of our initial posi-tion. We lacked any spirit . . . Sighted a "Josef Stalin 122" tank at 2000 meters. Fired eight anti-tank shells; their tracer light was wonderful to see. They just bounced off, though the gun fired won-derfully." In the evening, the assault guns were able to break into the village, where in the dark a "Stalin" tank could be shot down from 100 meters away, and exploded after being hit five times (!). This description of action is reminiscent of the first year of the eastern campaign, when assault artillerymen with their short 7.5 cm guns needed up to six shots to be able to put a T-34 tank out of action. All the same, the assault gun was indispensible for antitank action. Characteristic of the situation at that time is the evaluation by an officer of the Wehrmacht command staff, who reported in a memo on February 17, 1945: ". . . The only effective antitank defense was provided by the assault guns and the pursuit tanks, while the heavy antitank guns (motorized platoon) proved more and more to be insufficient."

Kinght's Cross bearer Alfred Regeniter, here still an Oberfähnrich, on his assault gun, photographed in the spring of 1944.

An Assault Gun III, Type G, of Assault Gun Brigade 303, which was in service with Army Unit Narva at Peipus Lake in the summer of 1944. It had been formed at Burg near Magdeburg on October 24, 1943. The troop emblem is easy to see on this vehicle.

This assault gun belonged to Assault Gun Brigade 667 in the spring of 1944. It already shows the return rollers with rubber-saving steel surfaces introduced in November 1943. The bow of the hull and the front of the upper body have been thickened to 80 mm with screwed-on armor plate. In addition, numerous spare tracks have been attached as makeshift protection. Note the artillery hit on the gun-cradle armor.

Assault Gun Brigade 667 had been organized at Jüterbog on July 24, 1942. It succeeded Assault Gun Battery 667 and used the unicorn as its troop emblem. In 1944 it was with the 9th Army in the east, in December 1944 it took part in the Ardennes offensive as Assault Artillery Brigade 667. On November 14, 1943 this unit recorded, in a Wehrmacht report, its 1000th tank shot down since 1942.

This assault gun, commanded by Knight's Cross bearer Oberleutnant Oberloskamp, Chief of the 1./ Assault Gun Brigade 667, at the barracks in Mogilyev, was shot down during the combat at Cherkassy on April 12, 1944, and later salvaged. The hull and upper body were hit hard by artillery shells.

In the course of the heavy fighting for the Allied beachhead in Anzio-Nettuno in the Italian theater of war in February 1944, no assault gun units saw action. The assault guns used here belonged to Panzer Unit 103 (3rd Panzergrenadier Division), Panzer Unit 129 (29th Panzergrenadier Division), and thus to the armored troops, as well as the "Hermann Goering" Armored Paratroop Division of the Luftwaffe.

The action of assault guns, as show here amid the ruins of the city of Cassino, did not correspond to the nature of this weapon, but it did strengthen the defensive power of Paratroop Jäger Regiment 3. In a strong Allied air raid on March 15, 1944, four of the five assault guns in the city were lost.

This assault gun was hit on the bow of the hull, the muzzle brake and the gun-cradle armor by British antitank gunfire. The picture was taken in Italy in the spring of 1944.

With the opening of the Allied front in Normandy on June 6, 1944, Assault Gun Brigades 341 and 394 of the Army were in France, as were Assault Gun Brigade XII of the Luftwaffe and SS Assault Gun Units 1 and 2 (1st and 2nd Panzer Divisions). Other assault guns were with several Panzer and Panzergrenadier divisions. The Allied air superiority considerably hindered the mobility of the assault gun units.

Of the 248 assault guns were reported in the west on June 10, 1944, most were lost. A great number of the losses were caused by vigorous fighter-bomber attacks. Assault gun Brigade XII brought back only one of its assault guns.

As of September 1944, parts of Assault Gun Brigade 280 were applied against the 1st British Paratroop Division, which had landed in Arnheim. The assault guns saw action in street fighting, individually and with the support of assault engineers. In one case, the leaderless crew of an assault gun disembarked after being fired on by six-pound (57 mm) antitank guns and abandoned their vehicle to the paratroops.

Assault Howitzer III, Type G (Sd.Kfz. 142/2), this one being an early version with boxlike gun-cradle armor and muzzle brake. Note the additional armor plate.

Assault Howitzer III, Type G (Sd.Kfz. 142/2), the type without a muzzle brake. It was omitted according to "Army Technical Order Sheet" No. 635 of September 1944. The barrel of the assault howitzer weighed 726 kilograms, 31 of which were eliminated with the muzzle brake.

An Assault Howitzer III of the 1./Assault Gun Unit 280 at Jüterbog. The unit operated on the eastern front as of the autumn of 1943. One year later it saw service in Arnheim and southwest of Tilburg, Holland. The "Zimmerit" protective coating is easy to see.

According to an experience report of the 29th Panzergrenadier Division of October 26, 1944 from Italy, "Assault guns are the infantry's best backbone on defense. Where they stand, the infantry also stands!" Panzer Unit 129 of this division had three assault gun companies.

Assault Gun IV of the assault gun company of a Panzerjäger unit in Greece. In 1944 these units, as parts of infantry divisions, usually included a company of Assault Gun III or IV.

The Assault Gun IV (Sd.Kfz. 167) had a fighting weight of 23 tons. Thanks to the use of the Panzer IV hull, the vehicle was somewhat roomier. 63 shells for the 7.5 cm Assault Cannon 40 could be carried. Here too, note the improvised strengthening of the 80 mm front armor with armor plates and concrete.

In this picture, the differences in the form of the armor plate on top of the upper body of the Assault Gun IV can be seen clearly. In front of the loader's hatch is the closed rack for the remote-control all-around machine gun and the opening for the close combat weapon. The supplying of the two weapons, so important for the assault guns, took place very slowly.

Assault Gun Unit 276 turns out at the troop training camp in Altengrabow in the summer of 1943. They saw service in the east with the Army Group Center.

The troop emblem of Assault Gun Unit 276.

Withdrawal across the Dniepr. The assault guns of the "Panther" Unit cross the river on a ferry.

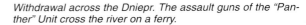

Assault guns of Assault Gun Unit 276 carrying infantrymen on the march. Between the self-propelled gun mounts is a "Maultier" tracked freight carrier (Sd.Kfz. 3) of the ammunition troop.

Note the camouflage paint which was used by Assault Gun Unit 276 in its first action.

This assault gun lost its track when a mine exploded in a minefield. To be able to repair the damage, the track aprons first had to be removed (note the bow of the hull). In this case, the attack had to be halted until engineers, under the cover of assault-gun fire, had cleared a lane.

Under heavy fire, an assault gun of the "Panther" Unit is supplied with ammunition during the winter of 1943-44.

At the end of January 1944, an assault gun of the 3./Assault Gun Unit 276 was fired on by a German tank platoon with 2 ch anti-aircraft guns in the vicinity of the Kovno Korosten rail line. One anti-tank shell penetrated the engine cover and damaged the powerplant.

Assault Gun Unit 276 being transported by rail. Lashed-down tarpaulins are meant to protect sensitive parts of the assault guns from dust and moisture.

This picture shows the 3./Assault Gun Unit 276 in transit in the east in February 1944.

An assault gun of the 3./Assault Gun Unit 276 broke through while crossing a bridge and turned over during an attempt to salvage it.

Preparations are made to tow an assault gun away. If another assault gun was used to tow it, then it was done by a rope, as long as possible, that had to be attached crosswise. This allowed better driving on curves. For self-towing, tree trunks and strong boards could be used.

The Assault Gun III, Type Gm with tactical number 333 belonged to the third platoon, third company of Assault Gun Unit 276. The commander's cupola does not yet have a steel deflector.

Towing tractors of the salvage troops tow an Assault Gun III away. In front is a 12-ton tractor (Sd.Kfz. 8), behind the 18-ton tractor (Sd.Kfz. 9) usually used for such jobs.

An assault howitzer of Assault Gun Unit 276.

As of August 1944, the newly organized Unit 276, now called an Assault Gun Brigade, saw service in East Prussia. Notice the tactical symbol on the track apron, used to differentiate the batteries the platoons.

According to the war strength schedule of June 1944, every assault gun battery was to include an escort grenadier battery. It was preferentially equipped with modern weapons, such as the assault gun (see picture), and was to be used primarily to take charge of securing the assault guns in all phases of combat and thus free the infantry of this task.

An Assault Gun III with all-around machine gun. This picture was taken on the eastern front in January 1945.

The Last Year of the War

In the last year of the war, the burning question for the assault artillery was how the growing tank fleets of the Allied armies could be held back. To do that, the Army's assault gun and assault artillery brigades needed assault guns with higher-performance weapons. The production of the Assault Gun III and IV was supposed to run out in August 1945. In place of them, a considerable increase in the delivery of the Jagdpanzer 38 "Hetzer" was planned. The total of 1300 vehicles was stated for the period from November 7, 1944 to July 1945. Series production of the Diesel-engined Jagdpanzer 38 was set for that month at five units. Production was to increase to 800 vehicles by December 1945. The development of a Diesel-engined pursuit tank came about under the pressure of increasingly poor gasoline supplies to the troops. On March 19, 1945, Guderian even requested an immediate change to Diesel engines. The Jagdpanzer 38 (Diesel) was to be armed with the 7.5 cm Panzerjägerkanone 39 L/48, the performance of which was no different from that of the 7.5 cm Sturmkanone 40 L/48. The installation of the 7.5 cm Panzerjägerkanone 42 L/70 was also planned. This weapon, using the 7.5 cm Panzergranate 39/42, could penetrate 500 mm of armor plate at 500 meters (60-degree striking angle). For the 7.5 cm Panzergranate 40/42 the penetration was 174 mm. Both versions of the Jagdpanzer 38 would have been the armored vehicles that, to an increasing degree, would have characterized the assault artillery units in the latter half of 1945. How unrealistic these plans were is shown by the fact that from January 1945 to the end of the war, only around 1200 Jagdpanzer 38 units were built (February: 401 + 5 repaired vehicles, March: 301 + 12 repaired). In the same period, 1133 Assault Gun III and IV units were delivered (February: 231 + 20 repaired, March: 322 + 26 repaired). The situation was also developing catastrophically for towing tractors; in all, 26 vehicles per month were reported as needed to equip the assault gun units. Less than 50 percent of the needs could be filled. A certain relaxation resulted here from the development and delivery of towing tanks.

In the spring of 1945, shortages typical of the fifth war year showed up more and more clearly in the assault gun and Army artillery brigades.

At the beginning of March 1945, the following numbers of assault guns were on hand: 3067 with 7.5 cm Sturmkanone 40 L/48, 577 with 10.5 cm Sturmhaubitze 42 L/28. The number of assault guns with the 7.5 cm Sturmkanone L/24, still 224, was significant.

After a German counterthrust in Goldap in mid-October 1944. Looking out the rear hatch of a "Panther" tank, we see a Panzerkampfwagen 38 (t) beside an assault gun.

The action strengths were low, and there were differences between the individual armies and army groups. This was particularly true of the 9th Army, which fought on the Oder front just outside the gates of Berlin. Its armored units were prepared for the expected large-scale Russian attack, the target of which was the German capital. Early in April 1945 two assault gun brigades were attached to this army. Assault Gun Brigade 111 had 42 assault guns and six Jagdpanzer IV/70; 37 of them were ready for action. Similarly high numbers could be reported by Assault Gun Brigade 210. Of its 41 combat vehicles, 13 of which were Jagdpanzer IV/70, 36 were ready for action. In all, there were 193 assault guns with the 9th Army, 164 of which could go into action immediately. The troops also awaited the arrival of sixteen new assault guns from the Berlin area.

It had become customary to apply the assault gun units some distance away from each other. Thus Assault Gun Brigade 210 had one battery in the Pomellen-Hohenholz area and two batteries in the Schwedt Forest (30 kilometers away).

On other sectors of the front, the commanders of the assault gun units could report such action figures in only a few cases. A few examples: On April 5, 1945, Assault Gun Brigade 202 had 20 Assault Gun IV and eight Jagdpanzer 38. Only eight Assault Gun IV and eight Jagdpanzer 38 were with Assault Gun Brigade 600 (both belonged to the Army Group Courland). From the Army Group E/Southeast Command came the strength report of the 3./Assault Gun Brigade 191: six Assault Gun III and two captured assault guns. The three assault gun brigades of with Army Group C/Southwest Command (No. 242, 907 and 914), with totals of 35, 33 and 25 assault guns, had considerable combat potential, though a large part was undergoing short- or long-term repairs. The equipment of Paratroop Assault Gun Brigade XXI consisted of a noteworthy 65 captured tanks. From the Army Group South came strength reports from Assault Gun Brigades 303 and 325 on April 1, 1945: eight and ten assault guns respectively. At this time, Assault Gun Brigade 256 (Army Group Center) had eight Jagdpanzer 38, and Brigade

280, with the Western Command had two Jagdpanzer IV and two Assault Gun III units. At this point the planned strengths of the assault gun brigades should be remembered: they were 31 or 45 assault guns. To be sure, there were limitations here too. In a message sent on February 17, 1945, Army Assault Gun Brigades 341 and 394 were ordered supplied with ten assault guns instead of the usual planned strength of fourteen per battery because of the major shortage. The offer was refused, though, by the Army High Command on February 21, 1945, but from the message it can be seen that a reduction in supplies had to be accepted sooner or later. The Army's armament study, published early in January 1945 under tit title "Assault Program", contained figures which indicated that only 900 assault guns, Jagdpanzer 38 and derivatives should be produced per month. Compared to the statistics in the earlier program, this was a considerable re-

duction, which must be regarded as being limited further as of February 1945. Too little material reached the assembly firms from their suppliers. In all, in mid-April 1945 there were 2084 assault guns with the troops, the greatest part of them in the assault gun brigades. It is striking that the equipping with various pursuit tanks, captured tanks and assault guns often differed from the war strength figures. The Jagdpanzer 38 in particular, with its narrow fighting compartment and limited field of fire, was hard for the assault artillerymen to get used to. Yet these vehicles saw action with the assault artillery in goodly numbers. In a message of March 2, 1945 (OBd.H./AHA/Stab Ia (2) No. 11874/45 geh.), it is said of the refreshing of Army Assault Artillery Brigade 236: "The Army Assault Artillery Brigade 236, assigned to the Training Command of Bohemia and Moravia in Milowitz, is to be equipped with Jagdpanzer 38 shortly."

A Panzerjäger 38 "Hetzer" that was abandoned in Bohemia early in May 1945. Of the 627 vehicles of this type that were available in the Wehrmacht at the beginning of April 1945, 502 were on the eastern front. In Courland at this time, Assault Gun Brigades 393, 600 and 912 were supplied with the "Hetzer". In Assault Gun Brigade 256 (Army Group Center) there were eight Panzerjäger 38 on March 31, 1945.

The action of the assault artillery suffered from further shortages in the spring of 1945. According to a notice from the Quartermaster General's Department of the Army General Staff dated January 14, 1945, a hitherto unknown decrease in ammunition supplying had to be expected in the coming spring. For the 7.5 cm tank and assault guns the ammunition supply would be decreased 62 percent from that of the autumn of 1944. The problem was the production of shells and cartridges. The fuel supply situation developed into something far more critical. The Army High Command informed the Army Group Vistula on March 31, 1945 that only 5 percent of the usual amount of gasoline and 10 percent of Diesel fuel would be available. It is not surprising that the regrouping suggested on April 7, 1945 was rejected, especially by the artillery of the army group. "Every drop of gasoline must be saved for the use of the tank units (and the assault artillery) in the great battle."

Losses of equipment in the assault gun units had increased markedly in the course of the retreat combat. Often the vehicles were abandoned with minor damage and could not be towed away for lack of towing vehicles. Thus they were counted among the total losses. 4490 assault guns were lost in 1944, 3468 of them on the eastern front, and in 1945 the production figures declined, so that a shortage of assault guns soon became evident, and the assault gun units no longer had any vehicles on hand as reserves. Typical of the situation in the last weeks of the war is a message of April 30, 1945, regarding the regrouping of assault artillery units into Panzerjäger units. It was stated: "Bands of the assault artillery that have lost their guns and that cannot be sent replacements in a short time are to be used as Panzerjäger units within the framework of the available close-combat means."

On December 21, 1944, the Red Army began the third battle of Courland with intensive fire from their artillery. Assault Gun Brigade 912, which saw service with its three batteries in the sector of the 11th, 205th and 290th Infantry Divisions, immediately moved out of the readiness area to make a counterattack. As usual, the three batteries had an area of 20 to 35 kilometers to defend. This makes clear what kinds of problems the assault gun brigades faced in the use and supplying of their batteries.

According to instructions from the war command of the 5th Panzer Army in the west on December 7, 1944, the infantry divisions were supposed to acquire as many assault guns as possible along the way as could be taken from, for instance, the armored divisions. Assault gun batteries were to be kept ready close behind the attacking spearheads. Experiences from the war year of 1941 were rediscovered in a Wehrmacht characterized by noticeable shortages.

These and many other difficulties accompanied the assault gun forces throughout the last weeks of World War II. It could be said in conclusion: If the assault gun proved itself to be an effective offensive weapon in the first years of the war, then it had to take over the protection of the infantry from the tank units to the extent that the tank, including those of the enemy, became the decisive weapon on the battlefield. In the course of the war, the conception of the assault gun changed into that of a tank destroyer. Despite great success in action, the assault gun forces had to come to grief with this task. The determination of victory or defeat in an event like World War II did not depend on the efficiency of a weapon, but rather to a much greater degree—and the combat history of the assault gun shows this very clearly—on the material and personnel resources of the states engaged in the war.

The troop emblem of Assault Gun Unit 277.

Assault guns of old production runs were still found individually in front service with the assault gun units in the last months of the war. In this picture we see one along with two Panzer IV tanks (7.5 cm Sd.Kfz. 161/2) in a counter-attack.

Such veterans came much more frequently from the assault gun replacement and training units, in whose work-shops they could at least be brought up to the current state of armament. As 1944 turned to 1945, such alarm batteries often saw service at the front; thus Assault Gun Replacement and Training Unit 200 applied six short-barreled assault guns in the Warthbrück area on January 12-13, 1945. This picture was taken near Altdamm in 1945 and shows an Assault Gun III, Type C, rearmed with the 7.5 cm Assault Cannon 40 L/48 and "sow's-head" shield.

This series of pictures shows a 7.5 cm Assault Gun, Type G, driving through a village in the Angermünde area as high speed on February 10, 1945. For the sake of the steering brakes, the top speed on suitable roads was not to exceed 22 kilometers per hour.

On April 10, 1945 there were 1053 Assault Gun III units and 198 assault howitzers with the troops at the front, with 812 and 157 of them on the eastern front, and only 54 and 41 in the west.

Assault Gun Brigade 243, under Hauptmann Rubig, was reorganized at the Jäger barracks in Potsdam in March and April 1945. Since there was a lack of assault guns at first, infantry service was planned for the 2nd and 3rd batteries. Then they presumably received the last assault guns that were completed by Alkett and sent directly to them, and at the beginning of their service with the 12th Army (Wenck) they had 35 assault guns and howitzers. The last three vehicles (without paint) were delivered on April 20, 1945.

Out of the shattered Units 278 and 322, Assault Gun Brigade 1170 was organized at the assault gun school in Burg. As of April 6, 1945, they had been sent 31 assault guns, 91 of which had to be picked up in Berlin-Spandau. As of April 13 the brigade was subordinate to Army High Command 12 and took part in the harsh combat southwest of Berlin. On April 28, 1945 it still had 14 assault guns ready for action.

An Assault Gun III, Type E, with the 7.5 cm Assault Cannon 40, seen in May 1945 after the end of street fighting in Berlin.

Assault Gun Brigade 276 was part of the 4th Panzer Division at the beginning of March 1945 and saw action in the Danzig area.

Shattered supply vehicles of an assault gun brigade. The tracked two-ton "Maultier" freight carriers (Sd.Kfz. 2) were utilized by the ammunition suppliers of the batteries. This picture was taken in the Danzig area in the spring of 1945.

These pictures show the service of Army Assault Artillery Brigade 311 in Lusatia at the end of April 1945. Here the brigade took part in the successful attacks on Bautzen and Weissenburg. By May 6, 1945 the unit had been loaded on trains at the depots in Dresden-Klotzsche and Radebeul and transferred to the Army Group Center. Parts of the brigade received news of the surrender at Teplitz-Schönau.

With the Army Group Center (Schörner), the members of Assault Gun Brigades 236, 300, 301 and 311 and the "Grossdeutschland" Assault Gun Brigade experienced the surrender of the German Wehrmacht in Bohemia and Moravia on May 8, 1945. Small groups from other units were with them as well.

A Jagdpanzer 38 "Hetzer" of Army Assault Artillery Brigade 236 in the vicinity of Teplitz-Schönau after the surrender.

German war materials became scrap iron; in the foreground is an assault gun (easy to identify by the gun mount and cradle armor), behind it are Jagdpanzer IV and 38.

In June 1949 this Assault Gun IV was blown up at Oderbruch, near the village of Gorgast, by workers of the Brandenburgische Bergungs AG. It had been lost in a minefield during combat in April 1945.

On February 15, 1986 this chassis of an Assault Gun III was towed out of an old branch of the Oder in the vicinity of Wriezen. In April 1945, the 25th Panzer Division was fighting in this area. According to a report from Army Group Vistula on April 6, 1945, that division had at that time, among others, 38 Assault Gun III units, 32 of which were ready for action.

In the years after the war, all the parts of the upper body that projected above the water were removed. The chassis and parts of the armament were salvaged only in 1985. Here the gun socket and the ZF SSG 77 Aphon gearbox, mounted at the side, can be seen.

A relic of the combat around the "Harz Fortress" in April 1945. The left side of the armored upper body of an Assault Gun III, which was salvaged in the vicinity of Schierke at the foot of the Brocken in May 1997.

This Assault Gun III was lost in Normandy in the summer of 1944. The armored upper body is now at the Museum of Military History in Dresden. Note that the upper front armor has been strengthened with concrete.

In 1943-44, Bulgaria had received a total of 55 German-made assault guns and organized two assault gun units with them. One of the vehicles is now at the Bulgarian Army Museum in Sofia.

Between June 1943 and August 1944, Finland received 59 assault guns and howitzers from Germany. Several of the vehicles delivered then are now in the German Tank Museum in Parola. Others were given to museums in other countries or, like the vehicle shown here, belong to private collectors (Bob Flaming, Great Britain).

This splendidly restored Assault Gun III, Type G, likewise came from Finland.

Today it belongs to the collection of the Museum of Military History in Dresden.

Front and rear views of the same assault gun. The two antenna mounts on the rear body suggest its early use as a command vehicle.

The following pictures were taken during a test of the assault gun on the grounds of the former vehicle testing facility in Kummersdorf.

Climbing over an obstacle.

The memory of the assault gun is linked with the recognition of a remarkable development of weapon technology in World War II. Similarly to the tank weapon, it has given that war a specific character.

Select Bibliography

H.Dv. 200/2 m	Ausbildungsvorschrift für die Artillerie, Vol. 2 m, Die Sturmgeschützbatterie, Berlin 1942.
H.Dv. 469/3 c	Panzerabwehr aller Waffen, Berlin 1942.
H.Dv. 481/57	Merkblatt für die Munition der 7,5-cm-Kampfwagenkanone und des Sturmgeschützes 7,5-cm-Kanonen Berlin 1940.
H.Dv. 481/58	Merkblatt für die Munition der 7,5-cm-Sturmkanone 40 und die 7,5-cm Kampfwagenkanone, Berlin 1944.
D 652/45	Sturmgeschütz 7,5-cm-Kanone, Ausführung A und B, vorläufige Justieranweisung. Berlin 1940 (reprint 1944).
D 693/9	Vorläufige Beschreibung und Umbauleitung des Funkgerätes in der gepanzerten Selbstfahrlafette für Sturmgeschütz (Ausführung E), Berlin 1942.
Merkheft	Gepanzerte Selbstfahrlafette für Sturmgeschütz 7,5-cm-Kanone (Sd.Kfz. 142), Merkheft für Kraftfahrzeugausbildung, Jüterbog 1943.
Merkblatt	Munitionsmerkblatt 3 (Sturmgeschütz), Berlin 1943.
Merkblatt	Merkblatt für den Wintereinsatz von Sturmgeschützen, Jüterbog 1943.
Merkblatt	Die Sturmbatterie-Einsatz- und Ausbildungsmöglichkeiten, zusammengestellt durch Offz. der VI. Lehrabteilung, no place or date.
Merkblatt-	Merkblätter für Artillerie Nr. 34 - Richtlinien für den Einsatz der Sturmgeschützeinheiten, Berlin 1942.
Merkblatt 18/10	Was muss der Grenadier vom Sturmgeschütz und Panzerjäger wissen? no place, 1943.
Geräteverzeichnis	Vorläufiges Geräteverzeichnis Kraftfahrgerät Teil 3 Fahrgestell Sturmgeschütz (7.5-cm)(Sd.Kfz. 142), Vol. 34, Berlin 1944.
Guderian, Heinz	Erinnerungen eines Soldaten, Stuttgart 1986.
Günter, H.	Erinnerungen an ernste und heitere Stunden der Sturmbrigade 277, no place or date.
Gruss, H.	Die deutschen Sturmbataillone im Weltkrieg, Berlin 1939.
Halder, F.	Kriegstagebuch, Vol. 2 and 3, Stuttgart 1963-64.
Kurowski, F.	Sturmartillerie, Stuttgart 1978.
Kröhne, W.	Tagebuch der Sturmgeschützbrigade 190, Düsseldorf 1955.
Lindenberg, G.	Die Abwehrkämpfe 1945 an der Oder und in Mecklenburg unter besonderer Berücksichtigung der Sturmgeschützbrigade 210, Stuttgart 1986.
Spielberger, W.J.	Sturmgeschütze, Stuttgart 1991.
Regeniter, A.	"Panzer auf 8 Uhr...", Rastatt, no date.
Thomas, F.	Sturmartillerie im Bild 1940-1945, Osnabrück 1986.
Ziesen, K.	Dreieinhalb Jahre bei der Sturmartillerie - Kriegstagebuch und Erinnerungen, Lüdenscheid, no date.